COMPANION IN DIRECT CURRENT TECHNOLOGY

Ashton Jairam

Prentice Hall
Upper Saddle River, New Jersey   Columbus, Ohio

Editor: Scott Sambucci
Production Editor: Rex Davidson
Design Coordinator: Karrie Converse-Jones
Cover Art and Design: Rod Harris
Production Manager: Patricia A. Tonneman
Marketing Manager: Ben Leonard

This book was printed and bound by Victor Graphics, Inc. The cover was printed by Victor Graphics, Inc.

Printed in the United States of America

10 9 8 7 6 5 4 3 2 1

ISBN: 0-13-080399-5

Prentice-Hall International (UK) Limited, *London*
Prentice-Hall of Australia Pty. Limited, *Sydney*
Prentice-Hall Canada Inc., *Toronto*
Prentice-Hall Hispanoamericana, S. A., *Mexico*
Prentice-Hall of India Private Limited, *New Delhi*
Prentice-Hall of Japan, Inc., *Tokyo*
Prentice-Hall (Singapore) Pte. Ltd., *Singapore*
Editora Prentice-Hall do Brasil, Ltda., *Rio de Janeiro*

PREFACE

In the teaching of direct current technology, I have found that most students find difficulty not in the basic theoretical principles but mainly in the application of the theoretical principles to the solution of practical problems.

This book attempts to bridge the gap between theoretical principles and practical application by presenting carefully detailed solutions to selected problems and, in some cases, showing alternative methods of approach.

Preceding the examples in each chapter, brief theoretical notes have been introduced, focusing on specific aspects of the subject matter only; no attempt has been made to produce a conventional type textbook but rather a companion sequence providing a guide to the methods and procedures in the solution to examination problems and for comparison with one's own personal efforts. Each chapter concludes with a collection of unsolved problems taken from past examination papers.

The range of problems provided covers most of the syllabuses in direct current technology programs conducted by Technical Institutes, Technical Colleges, Colleges of Applied Arts and Technology and Polytechnics at the introductory, advanced undergraduate and graduate levels. This apart, this book may be considered a suitable companion in continuing education and supplemental studies.

Most of the problems have been selected in part from past papers of examining bodies, and in this connection I wish to express my gratitude to the authorities for using extracts from their past examination papers.

Moreover, I wish to express my appreciation to the final year students of the electrical engineering technician program of the Gulf Polytechnic, Isa Town, Bahrain, Arabian Gulf, the final year electrical technician engineering students of the Government Technical Institute, Georgetown, Guyana, and the graduate technician engineers in training at the G.E.C. Training Department, Sophia, Georgetown, Guyana, for their tireless efforts in researching sources and in finding solutions to many of the past examination papers.

It is hoped that this book will provide direction to students in Technical Institutes and those working on their own to broaden their horizons in direct current technology.

A.J.

ABOUT THE AUTHOR

Born in Guyana, educated and trained in the U.K., and now living with his wife Barbara in Toronto, Canada, Ashton Jairam is a Chartered Electrical Engineer and administrator with an international background of diversified experience in electronics and the electric utility industry, with special emphasis on Human Resources development. Ashton has been involved in professional training and development in both electronics and electric utility services for more than 20 years. His experience includes Training Manager G.E.C. Training Department; Senior Lecturer, Gulf Polytechnic and Guyana Government Technical Institute; Product Supervision Engineer, Smiths Industries; and Technical Adviser and Chief Design Engineer, Crossfield Electronics. He is currently involved in system analysis programs, personal computers, and Human Resources development and uses both MS-DOS and Microsoft Windows to write a wide range of simple modular training programs supported by CorelDRAW for graphic production.

D.J.L.

ABOUT THE BOOK

This book is the third in a major series of three books covering the fundamental aspects of electrical technology. The author's carefully chosen examples demonstrate the application of theory to practical problems. The Companion Series should prove to be a valuable source of reference for students and will be retained by them through their technician and engineering careers.

D.J. Longman O.B.E.
B.Sc.Eng. (London), F.I.C.E.
M.I.Struct.E, C.Eng.

CONTENTS

I often say that when you can measure what you
are speaking about, and express it in numbers, you
know something about it; but when you cannot
express it in numbers, your knowledge is of a meager
and unsatisfactory kind; it may be the beginning of
knowledge, but you have scarcely, in your thoughts,
advanced to the stage of science, whatever the matter
may be.

LORD KELVIN

CHAPTER   1

UNITS – ELECTRICAL AND MECHANICAL RELATIONSHIPS

Ohm's Law; Resistivity; Temperature Coefficient of Resistance;

Heat Energy and Power

Introduction

Ohm's Law

The relationship between an electrical current I amperes flowing in a linear conductor and the potential difference (p.d) in volts is expressed by Ohm's Law, which states that the p.d between any two fixed points on the conductor is directly proportional to the current flowing in the conductor, provided all other physical factors remain unchanged. Expressed thus

$$V \propto I$$
$$\therefore \quad V = I R$$

where   V   = potential difference in volts

   I   = current in the conductor in amperes

   R   = resistance of the conductor in ohms

The Coulomb – Unit of Quantity of Electricity

The unit of quantity of electricity (Q) is the coulomb. The rate of flow in coulombs/sec is the current I amperes. It is defined as the quantity of electricity in coulombs which passes a given point in an electric circuit in 1 second. Thus

$$Q = \text{amperes x time}$$
$$= I \text{ x } t \text{ coulombs}$$

or $I = Q/t$ amperes

Energy – Unit, Joule (J)

The force required to move a quantity of electricity completely around an electrical circuit is called the electromotive force (e.m.f, symbol E) and its unit of measurement is the volt (same as p.d). Energy is expended or work done by a force in moving coulombs in the circuit. Hence

Energy (joules)   = electromotive force x coulombs

$$= E \text{ x } Q$$
$$= E \text{ x } I \text{ x } t \quad (Q = It)$$

The Joule (Mechanical Relationship)

The joule is the work done by a force of 1 newton acting over a distance of 1 meter in the direction of the force.

$$1 \text{ Joule} = 1 \text{ newton meter}$$

Power

Power is defined as the rate of energy dissipation or rate of doing work and is expressed as

$$\text{Power (P) in watts (W)} = \frac{\text{energy (J)}}{\text{time (sec)}}$$

$$\textbf{1 Watt} = \textbf{1 joule/sec}$$

$$\text{Energy in joules (J)} = \text{power (P) in watts (W) x time (sec)}$$

The units of electrical energy and mechanical energy are related thus

$$1 \text{ J} = 0.737 \text{ ft-lbf}$$

$$\text{or} \quad 1 \text{ ft-lbf} = 1.357 \text{ J}$$

Horsepower (h.p)

The unit of electrical power is related to the unit of mechanical power

$$\textbf{1 Horsepower} = \textbf{746 watts} = \textbf{33000 ft-lbf/min} = \textbf{550 ft-lbf/sec}$$

Force – Unit, Newton

The force required to give a mass of 1 kg an acceleration (a) of 1 meter per second per second (m/sec^2) is called the newton.

$$\text{Force (N)} = \text{mass (kg) x acceleration } (a) \text{ (m/sec}^2)$$

Gravitational Force

The weight of a body is the gravitational force acting on the mass of that body. For almost all practical purposes, 9.81 N is considered to be the weight of a mass of 1 kg. That is

$$\text{Weight of 1 kg} = 9.81 \text{ N}$$

Potential Energy (P.E)

This is the energy acquired by virtue of the position of a body relative to earth and within the influence of gravitational force. For example, if a mass M kg is lifted vertically through a height of h meters, the potential energy (P.E) acquired by virtue of its position is given by

$$\text{P.E} = \text{mass x height x gravitational force}$$

$$= \text{M x h x 9.81}$$

$$= 9.81 \text{ Mh joules}$$

Kinetic Energy (K.E)

This is energy in motion. When potential energy (P.E) is released or set in motion, moving at a velocity of v meters/sec, it converts into kinetic energy (K.E) expressed as

$$\text{K.E} = \tfrac{1}{2} M v^2 \quad \text{joules}$$

Relationship between Displacement, Velocity, Acceleration, and Time

The formulae are important and collected together for convenience.

Let s = distance traveled in meters (m)

t = time in seconds (sec)

u = initial velocity in meters

v = final velocity in meters/sec after t seconds

f = retardation

a = acceleration (m/sec^2)

g = 9.81 m/sec^2 (acceleration due to gravity)

Uniform velocity

$$a = 0$$
$$s = v\,t$$

Uniform acceleration

$$v = u + at$$
$$s = \frac{(u + v)t}{2}$$
$$v^2 - u^2 = 2as$$
$$s = ut + \tfrac{1}{2}at^2 \quad (a+ \text{ for accelerating; } a- \text{ for retarding})$$

Falling body - Downward direction (that is, gravitational acceleration will be treated as negative)

$$v = u - gt$$
$$v^2 - u^2 = -2gs$$
$$s = ut - \tfrac{1}{2}gt^2$$

Work done – unit, joule (see definition of joule)

Work is done when a force of 1 newton moves a distance of 1 meter in the direction of the force. That is

$$\text{Work done (J)} = \text{force (N) x distance (m)} \quad \text{N m}$$
$$\textbf{1 Joule} = \textbf{1 N m (newton meter)}$$

Resistance and Resistivity

The resistance R of a conductor is proportional to its length and inversely proportional to its area. That is

$$R \propto l$$
$$R \propto 1/a$$
$$\therefore \quad R = \rho \, l/a \quad \text{ohms} \qquad \text{(Eq. 1.0-0)}$$

where ρ = resistivity of the material, or resistivity (Ω-m)

l = length in meters

a = area in square meters (m^2)

Resistivity (ρ) is defined as the resistance between opposite faces of a unit cube at a given temperature.

Temperature Coefficient of Resistance (α)

This is defined as the change in resistance of a conductor per 1°C change in temperature expressed as a fraction of its resistance at 0°C. The temperature coefficient of resistance is denoted by α. For example, the resistances R_1 (Ω) at $t_1{}^{\circ}$C and R_2 (Ω) at $t_2{}^{\circ}$C for a conductor with R_o ohms at 0°C a temperature coefficient α_o at 0°C can be expressed as follows:

$$R_1 = R_o (1 + \alpha_o t_1) \qquad \text{(Eq. 1.0-1)}$$
$$R_2 = R_o (1 + \alpha_o t_2) \qquad \text{(Eq. 1.0-2)}$$

Dividing Eq.1.0-1 by Eq.1.0-2, we get

$$\frac{R_1}{R_2} = \frac{R_o (1 + \alpha_o t_1)}{R_o (1 + \alpha_o t_2)} \qquad \text{(Eq. 1.0-3)}$$

In general, if a material has a resistance R_{20} at 20°C and α_{20} at 20°C, the resistance R_{t2} at a temperature t_2 ($t_2 > 20^{\circ}$C) is given by

$$R_{t2} = R_{20} [1 + \alpha_{20} (t_2 - 20)] \; \Omega \quad \text{(Eq. 1.0-4)}$$

Specific Heat (J/kg.K)

The specific heat of a substance is the heat required to raise the temperature of 1 kg of that substance through 1°C. The unit of specific heat is J/kg.K.

[Note: The heat required to raise 1 g of water by 1°C is 1 cal; if the specific heat is quoted in calories/gram, then to convert calories to joules multiply by 4.18. A temperature rise of 1°C = a temperature rise of 1 K.]

(contd)

Assume one wishes to boil a quantity of water in an aluminum pot weighing 1.2 kg. Energy will be absorbed in heating the aluminum pot in addition to boiling the water. The quantity of heat utilized to raise the temperature of the aluminum pot from ambient temperature (say 20^oC) to 100^oC is given by the expression

Quantity of heat = mass x specific heat x temperature rise

If the specific heat of aluminum is taken to be 950 J/kg.K, then

Quantity of heat = 1.2 (kg) x 950 x (100 – 20) = 91200 J

This quantity of heat (91200 J) is called the *water equivalent of the aluminum pot* and must be added to the heat used in boiling the water to obtain the total amount of heat consumed.

Efficiency (η)

Efficiency of a piece of equipment may be defined as its ability to perform well or its quality of being efficient. It is usually expressed as a percentage of the ratio of the equipment's output to its input. That is

$$\text{Efficiency } (\eta) = \frac{\text{output}}{\text{input}} \times 100\%$$

$$= \frac{\text{input} - \text{losses}}{\text{input}} \times 100\%$$

$$= 1 - \frac{\text{losses}}{\text{input}} \times 100\%$$

Units

Measurement	Unit
length	meter (m)
mass	kilogram (kg)
time	second (sec or s)
current	ampere (A)
potential difference (p.d)	volt (V)
electromotive force (e.m.f)	volt (V)
resistance (R)	ohm (Ω)
power (P)	watt (W)
temperature	kelvin (K)

<u>Common conversion factors</u>

1 in	= 2.54 cm	1 hp	= 746 W
1 N	= 0.225 lbf	1 hp	= 550 ft-lbf/sec
1 kgf	= 9.81 N	1 cal	= 4.187 J
1 lb	= 453.6 g	1 BTU	= 252 cal
	$\approx$ 0.454 kg		= 1055 J
1 J	= 0.737 ft-lbf	1 N m	= 1 J
1 kW	= 1 J/sec	1 kWh	= 3.6 x 10^6 J

1 gal (imperial measure) of pure water is taken to weigh 10 lb

<u>Specific heats for some well-known materials between 0ºC and 100ºC</u>

<u>Material</u>	<u>Specific Heat</u>
Copper	390 J/kg.K
Iron	500 J/kg.K
Aluminum	950 J/kg.K

Specific heat of pure water is taken as 4190 J/kg.K

Example 1.1

Determine the quantity of electricity in coulombs when 10 A passes a given point in the circuit for 5 min.

Solution

$$\text{Quantity, } Q = I \text{ (amperes) x } t \text{ (sec)}$$
$$= 10 \text{ x } 5 \text{ x } 60 \qquad = \textbf{\underline{3000 coulombs}}$$

Example 1.2

Calculate the current flowing in a circuit in which 200 coulombs passes a given point in 10 sec.

Solution

$$Q = I\,t$$
$$\therefore \quad I = Q/t$$
$$= {}^{200}/_{10} \qquad = \textbf{\underline{20 A}}$$

Example 1.3

Calculate the time required for 15,000 coulombs to pass a given point in a circuit when the current is 25 A.

Solution

$$Q = I\,t$$
$$\therefore \quad t = Q/I$$
$$= 15000/25 \qquad = \underline{\textbf{600 sec}}$$
$$= \underline{\textbf{(10 min)}}$$

Example 1.4

Given that

$$1\ lb\ = 453.6\ g, \quad 1\ in\ = 2.54\ cm,$$
$$1\ h.p\ = 550\ ft\text{-}lbf/sec \quad and \quad 1\ BTU\ = 1055\ J$$

calculate

    (i)   *how many joules are equivalent to 1 ft-lb*

    (ii)   *how many watts are equivalent to 1 h.p*

    (iii)   *how many ft-lb are equivalent to 1 BTU*

Solution

(i)

$$1\ J = 1\ N\,m$$
$$1\ ft = 12\ in$$
$$1\ lb = 453.6\ g$$

Converting feet to meters, we get

$$1\ ft = \frac{12\ (in) \times 2.54\ (cm)}{100\ (cm)}\ m \qquad = \underline{0.3048\ m}$$

Converting pounds to newtons, we get

$$1\ lb = \frac{453.6\ (kg) \times 9.81\ N}{1000} \qquad = \underline{4.45\ N}$$

$$\therefore \quad 1\ ft\text{-}lb = 4.45 \times 0.3048 \qquad = \underline{\textbf{1.356 N m (joules}})$$

(ii)

$$1\ h.p = 550\ ft\text{-}lbf/sec$$
$$1\ ft\text{-}lb = 1.356\ N\,m$$
$$\therefore \quad 1\ h.p = 550 \times 1.356 \qquad = \underline{746\ N\ m/sec}$$

Now $1\ N\,m = 1\ J$

$$\therefore \quad 746\ N\ m/sec = 746\ J/sec \qquad = \underline{\textbf{746 W}}$$

(contd)

(iii)

$$1 \text{ BTU} = 1055 \text{ J}$$

$$1.356 \text{ N m} = 1 \text{ ft-lb}$$

$$\text{Now} \quad 1 \text{ N m} = 1 \text{ J}$$

$$\therefore \quad 1 \text{ BTU} = \frac{1055}{1.356} \qquad = \underline{\textbf{778 ft-lb}}$$

Alternative method

When in doubt of the final unit, a method known as **'dimensions'** is used to obtain the final unit of measurement. In this method, all numerics are accompanied by their associated units in the equation. The procedure is to cancel all common units and numerics; the remainder is the final unit of measurement.

(i)   To convert feet to meters and pounds to newtons

$$1 \text{ ft-lb} = \left[\frac{12 \text{ in}}{1 \text{ ft}} \times \frac{2.54 \text{ cm}}{1 \text{ in}} \times \frac{1 \text{ m}}{100 \text{ cm}} \right] \left[\frac{453.6 \text{ g}}{1 \text{ lb}} \times \frac{1 \text{ kg}}{1000 \text{ g}} \times \frac{9.81 \text{ N}}{1 \text{ kg}} \right]$$

$$= \frac{(12 \times 2.54) \text{ m}}{100} \times \frac{(453.6 \times 1 \times 9.81) \text{ N}}{1000} \quad \begin{array}{l} = \underline{\textbf{1.356 N m}} \\ \text{or} \; = \underline{\textbf{1.356 J}} \end{array}$$

(ii)   Similarly, to convert watts to h.p

$$1 \text{ h·p} = \frac{550 \text{ ft-lb}}{\text{sec}} \times \frac{1.356 \text{ J}}{1 \text{ ft-lb}}$$

$$= 550 \times 1.356 \text{ J/sec} \quad = \underline{\textbf{746 W}}$$

(iii)   Finally, to convert ft-lb to BTU

$$1 \text{ BTU} = 1055 \text{ J}$$

$$= 1055 \text{ J} \times \frac{1 \text{ ft-lb}}{1.356 \text{ J}}$$

$$= \underline{\textbf{778 ft-lb}}$$

Example 1.5

A train traveling between two stations 6 km apart completes the journey in 10 min. During the first 60 sec the train moves with constant acceleration. At the end of the journey the train comes to a standstill after braking for 40 sec. Between the first minute and the last 40 sec of the journey, the train traveled at uniform speed. Calculate the values of (a) the uniform speed, (b) the acceleration, and (c) the retardation.

(contd)
Solution

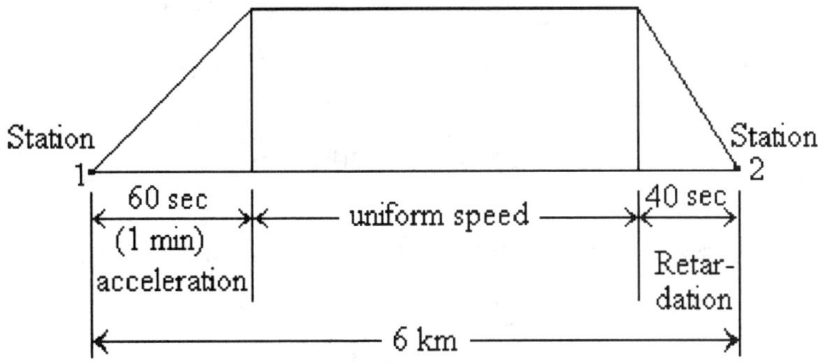

Fig. 1.5-0. Diagrammatic representation of the train's journey.

<u>Refer to Fig. 1.5-0</u>

Initial velocity, u = 0

For the first 60 sec, v = acceleration x time

$$= a \times 60 \qquad = \underline{60\,a} \qquad \text{(Eq. 1.5-0)}$$

During the last 40 sec, u = velocity – (retardation x time)

$$0 = v - f\,t$$
$$= v - 40\,f$$
$$\therefore \qquad v = \underline{40\,f} \qquad \text{(Eq. 1.5-1)}$$

Equating Eqs. 1.5-0 and 1.5-1, we get

$$60\,a = 40\,f$$
$$\therefore \qquad f = \frac{60\,a}{40} \qquad = \underline{1.5\,a} \quad \text{(Eq. 1.5-2)}$$

Distance traveled during acceleration is

$$s = ut + \tfrac{1}{2}at^2$$
$$= 0 + \tfrac{1}{2}\,a \times 60^2 \qquad = \underline{1800\,a}\ \text{m}$$

From Eq. 1.5-0 $\qquad v = a\,t$

$$a = {}^{v}\!/t$$
$$= {}^{v}\!/60$$
$$\therefore \qquad s = 1800\ \text{m} \times {}^{v}\!/60 \qquad = \underline{30\,v}\ \text{m}$$

Distance traveled at uniform speed is

$$s = \text{velocity x time taken to complete}$$

journey (i.e., acceleration time + retardation time)

$$= v\,t = v\,[(10 \times 60) - (60 + 40)]$$
$$= \underline{500\,v}\ \text{m}$$

(contd)

Distance traveled during retardation is

$$s = vt - \tfrac{1}{2}ft^2 \quad \text{(negative sign indicates retardation)}$$
$$= 40v - \tfrac{1}{2}f \times 40^2$$

From Eq. 1.5-1 $\qquad f = {}^{v}/40$

$$\therefore \quad s = 40v - \frac{800v}{40} = \underline{20\,v\ m}$$

Total travel distance, 6 km = acceleration + uniform speed + retardation

That is $\quad$ 6000 m $= 30v + 500v + 20v = \underline{550v}$

$$\therefore \quad v = \frac{6000}{550} = \underline{10.91\ m/s}$$

(a) **Uniform velocity,** $v = \underline{\textbf{10.91 m/s}}$

(b) **Acceleration,** $a = {}^{v}/60 \qquad$ [from Eq. 1.5-0]

$$= {}^{10.91}/60 \qquad\qquad = \underline{\textbf{0.182 m/s}^2}$$

(c) **Retardation,** $f = {}^{v}/40 \qquad$ [from Eq. 1.5-1]

$$= {}^{10.91}/40 \qquad\qquad = \underline{\textbf{0.273 m/s}^2}$$

Example 1.6

An electric kettle with an element rated at 250 V, 1 kW is filled with 3 pints of water at 15 °C. It is required to raise the temperature of the water to the boiling point (100 °C). The efficiency of the kettle is 91.6%. Calculate

(a) *the resistance of the heating element*

(b) *the cost of energy consumption at 20 cents per kWh*

Assume 1 gal of water to weigh 10 lb and the specific heat of water to be 4190 J/kg.K; 1 kg = 2.2 lb.

Solution

(a) $\qquad$ Power = V x I $\quad$ and $\quad$ I = ${}^{V}/R$

$$\therefore \quad \text{Power} = V \times {}^{V}/R = V^2/R$$

$$\therefore \quad R = \frac{V^2}{\text{power}} = \frac{250^2}{1000} = \underline{\textbf{62.5 }\Omega}$$

(contd)

(b)

Mass of water	$= {}^3/8 \times 10$	(3 pints $= {}^3/8$ of a gallon)
(Convert to kg)	$= ({}^3/8 \times 10) \times {}^1/2.2$	$= \underline{1.7\ kg}$
Temperature rise	$= (100 - 15)^{\circ}C$	$= \underline{85^{\circ}C\ (85\ K)}$

Quantity of heat required to boil water is

Useful heat $=$ mass x specific heat x temperature rise

$\qquad\qquad\quad = 1.7 \times 4190 \times 85 \qquad = \underline{605455\ J}$

$$\text{Efficiency, } \eta = \frac{\text{heat output}}{\text{heat input}}$$

$$\text{Heat input} = \frac{\text{heat output}}{\text{efficiency}}$$

$$= \frac{605455 \times 100}{91.6} \qquad = \underline{660977\ J}$$

[*Note: The difference between the input heat and output heat is the heat loss toward heating the kettle.*]

$$1\ kWh = 1000 \text{ watt hour}$$
$$= 1000 \times 3600 \text{ watt seconds or joules}$$
$$= 3,600,000\ J \quad (3.6 \times 10^6\ J)$$

$$\text{Heat input expressed in kWh} = \frac{660977}{3.6 \times 10^6} \qquad = \underline{0.1836\ kWh}$$

$$\text{Cost of energy} = 0.1836 \times 20 \qquad = \underline{\textbf{3.67 cents}}$$

Example 1.7

A steam power station has an overall efficiency of 25% and uses residual fuel oil with a calorific value of 6.2 x 10^6 BTU per barrel. Calculate the daily (24 h) consumption of fuel in gallons when the average load on the station is 60 MW. [Take 1 barrel of fuel to be 42 gal]

Solution

$$\text{Power output} = 60\ MW\ (MJ/s)$$

$$\text{Power input} = \frac{\text{output}}{\text{efficiency}}$$

$$= 60\ (MJ/s) \times \frac{100}{25} \qquad = \underline{240\ MJ/s}$$

$$\text{Energy input per day} = \text{joules x hours/day x time (sec)}$$
$$= 240 \times 10^6 \times 24 \times 60 \times 60$$
$$= 24 \times 24 \times 36 \times 10^9 \quad = \underline{20736 \times 10^9\ J/day}$$

(contd)

$$\text{Now 1 BTU} = 1055 \text{ J}$$

$$\therefore \quad \text{BTU/day} = \frac{20736 \times 10^9}{1055} = \underline{19.65 \times 10^9 \text{ BTU/day}}$$

Fuel consumption in barrels/day is

$$= \frac{19.65 \times 10^9}{6.2 \times 10^6} = \underline{3.169 \times 10^3 \text{ barrel/day}}$$

Fuel consumption in gallons/day is

$$= 3.169 \times 10^3 \times 42 = \underline{\textbf{133} \times \textbf{10}^\textbf{3} \textbf{ gal/day}}$$

Example 1.8

A crane raises a load of 2000 kg vertically at a speed of 18 m/min.
If the overall efficiency of the motor and gears is 56 % and the supply is
200 V d.c, calculate the current taken by the motor.

Solution

Power required by crane to raise load is

$$= \frac{\text{mass x distance}}{\text{time}}$$

$$= 2000 \text{ (kg)} \times 18 \text{ m/min}$$

$$= \underline{36 \times 10^3 \text{ kg m/min}}$$

$$\text{Power input to motor} = \frac{\text{output}}{\text{efficiency}}$$

$$= \frac{36 \times 10^3}{0.56} = \underline{64286 \text{ kg m/min}}$$

Power input to motor in watts (convert kg m/min to watts) is

$$\text{Power (watts)} = \frac{64286 \text{ kg m}}{\text{min}} \times \frac{1 \text{ min}}{60 \text{ sec}} \times \frac{9.81 \text{ N}}{1 \text{ kg}}$$

$$= \frac{64.286 \times 10^3 \times 9.81}{60} \text{ N m/sec (J/s)}$$

$$= \underline{10.51 \times 10^3 \text{ W}}$$

Current taken by motor is

$$I = \frac{\text{power}}{\text{voltage}}$$

$$= \frac{10.51 \times 10^3}{200} = \underline{\textbf{52.55 A}}$$

Example 1.9

A steam power station uses coal of a calorific value of 12000 BTU/lb
and has an overall efficiency of 23%. Calculate the weight of coal used
per kilowatt-hour generated when the average load on the station is
100 MW. [Take 1 cal = 4.18 J; 1 BTU = 252 cal]

Solution

$$\text{Power} = \frac{\text{output}}{\text{efficiency}}$$

$$= \frac{100}{0.23} \qquad = \underline{434.78 \text{ MW (MJ/s)}}$$

$$\text{Energy input/day} = \frac{\text{joules}}{\text{sec}} \times \text{time (sec)}$$

$$= 434.78 \times 10^6 \times 24 \times 3600$$

$$= \underline{375650 \times 10^8 \text{ J/day}}$$

$$= \underline{37.565 \times 10^{12} \text{ J/day}}$$

(Now 1 cal = 4.18 J and 1 BTU = 252 cal)
Therefore energy input in BTU per day is

$$\text{BTU/day} = \frac{37.565 \times 10^{12}}{4.18 \times 252} \qquad = \underline{35662 \times 10^6 \text{ BTU/day}}$$

Calorific value of coal is = 12000 BTU/lb

$$\text{Coal consumption (lb/day)} = \frac{35662 \times 10^6}{12 \times 10^3} \qquad = \underline{2971.83 \times 10^3 \text{ lb/day}}$$

$$\text{Coal consumption (lb/kWh)} = \frac{2971.83 \times 10^3}{\text{kW output} \times \text{hr}}$$

$$= \frac{2971.83 \times 10^3}{100 \times 10^3 \times 24} \qquad = \mathbf{\underline{1.238 \text{ lb/kWh}}}$$

Example 1.10

A train having a mass of 300 Mg is pulled by its engine at a constant
speed of 90 km/h along a straight horizontal track. The track resistance is
5 mN (5 x 10^{-3} N) per newton of train weight. Calculate (a) the tractive
effort in kilonewtons, (b) the energy in megajoules and kWh expended in
10 min, (c) the power in kW, and (d) the kinetic energy of the train in
kWh (neglecting rotational inertia).

(contd)

Solution

(a)

$$\text{Mass of train in kg} = \frac{300 \times 10^6}{10^3} = \underline{300 \times 10^3 \text{ kg}}$$

$$\text{Now the weight of 1 kg mass} = 9.81 \text{ N}$$

$$\text{Weight of train in newtons} = \frac{300 \times 10^3 \times 9.81}{10^3} \text{ kN}$$

$$\text{Tractive resistance} = 5 \times 10^{-3} \text{ N per newton of train weight}$$

$$\therefore \quad \text{Tractive effort} = 5 \times 10^{-3} (300 \times 9.81) \text{ kN}$$

$$= \underline{\mathbf{14.715 \text{ kN}}}$$

(b)

$$\text{Energy expended} = \frac{\text{force} \times \text{distance moved}}{\text{time}}$$

$$= \frac{14.715 \times 10^3 \times 90 \times 10^3 \times 10 \times 60}{3600} \text{ J}$$

$$= \underline{\mathbf{220.725 \text{ MJ}}}$$

$$1 \text{ kWh} = 3.6 \times 10^6 \text{ J}$$

$$\text{Energy expended in kWh} = \frac{220.725 \times 10^6}{3.6 \times 10^6} = \underline{\mathbf{61.3125 \text{ kWh}}}$$

(c)

$$\text{Power in kW} = \frac{\text{force} \times \text{distance}}{\text{time}} \times 10^{-3} \text{ kW}$$

$$= \frac{14.715 \times 10^3 \times 90 \times 10^3}{3600} \times 10^{-3} \text{ kW}$$

$$= 0.3679 \times 10^3 \text{ kW} = \underline{\mathbf{367.9 \text{ kW}}}$$

(d)

$$\text{Speed of 90 km/h} = \frac{90 \times 10^3}{3600} = \underline{25 \text{ m/s}}$$

$$\text{Kinetic energy (K.E)} = \tfrac{1}{2} M v^2$$

$$= \tfrac{1}{2} \times 300 \times 10^3 \text{ kg} \times 25^2$$

$$\text{K.E in kWh} = \frac{\tfrac{1}{2} \times 300 \times 10^3 \times 25^2}{3.6 \times 10^6}$$

$$= \frac{150 \times 10^3 \times 625}{3.6 \times 10^6}$$

$$= \frac{150 \times 625}{3600} = \underline{\mathbf{26.04 \text{ kWh}}}$$

Example 1.11

*An electric kettle takes 3.8 A from a 230 V supply. Determine the time
taken to raise the temperature of 3 pints of water from 12°C to 90 °C
and the cost of energy used at 15 cents per kWh. Assume 75%
of the energy to be usefully employed. [Take 4.18 J = 1 cal]*

Solution

$$
\begin{aligned}
\text{Weight of 1 gal water} &= 10\ \text{lb} \\
\text{Weight of 1 pint water} &= {}^{10}/_{8}\ \text{lb} = 1.25\ \text{lb} \\
1\ \text{lb} &= 453.6\ \text{g} \\
\text{Weight (grams) of 3 pints water} &= 3 \times 1.25 \times 453.6 = \underline{1701\ \text{g}} \\
\text{Temperature rise} &= (90 - 12)°\text{C} = \underline{78\ °\text{C}} \\
\text{Heat (useful) given to water} &= \text{mass of water} \times \text{temperature rise} \\
&= 1701 \times 78 = \underline{132678\ \text{cal}} \\
\text{Useful heat} &= 75\ \% = 132678\ \text{cal} \\
\text{Actual heat generated} &= 132678 \times \frac{100}{75} = \underline{176904\ \text{cal}} \\
\text{Power consumption of kettle} &= \text{V} \times \text{I} = 230 \times 3.8 = \underline{874\ \text{W}} \\
\text{Energy in joules/sec} &= 874\ \text{W} = \underline{874\ \text{J/s}} \\
\text{Time needed to raise temperature} &= \frac{176904 \times 4.18\ \text{J}}{874\ \text{J/s}} = \mathbf{\underline{846\ sec\ (14.1\ min)}} \\
\text{Energy consumed in kWh} &= \frac{176904 \times 4.18\ \text{J}}{3.6 \times 10^{6}} = \underline{0.2054\ \text{kWh}} \\
\text{Cost of electrical energy} &= 0.2054 \times 15 = \mathbf{\underline{3.08\ cents}}
\end{aligned}
$$

Example 1.12

*A copper conductor carrying a current of 2000 A is 518.16 cm long
with an internal diameter of 3.81 cm and an external diameter of 5.08 cm.
The voltage drop in the conductor is 0.2 V. Calculate the resistivity of the
conductor.*

Solution

$$
\begin{aligned}
\text{Resistance of conductor} &= {}^{\text{V}}/_{\text{I}} = {}^{0.2}/_{2000} = \underline{1.0 \times 10^{-4}\ \Omega} \\
\text{Cross-sectional area of conductor} &= {}^{(\pi d^{2})}/_{4} = 3.142\,(5.08^{2} - 3.81^{2})/_{4} \\
&= 3.142\,(25.81 - 14.52)/_{4} = \underline{8.868\ \text{cm}^{2}} \\
\text{From}\ \ R = \rho\,l/a, \quad \rho = R\,a/l &= \frac{1 \times 10^{-4} \times 8.868}{518.16} = \mathbf{\underline{1.71\ \mu\Omega\text{-cm}}}
\end{aligned}
$$

Example 1.13

*The reservoir of a hydroelectric power station is situated at a height of
150 m above the turbine driving the generators. Assuming that the
station has an overall efficiency of 75%, calculate the rate of flow of
water in gallons per minute through the turbines when the station is
developing 50 MW. [1 gal of water weighs 4.54 kg]*

Solution

The water flowing down from the reservoir possesses kinetic energy

Kinetic energy (K.E) $= \frac{1}{2} Mv^2$

The initial velocity $u = 0$; $s = -150$ m

Striking velocity (v) of the water at the turbine blades is obtained from

$$v^2 - u^2 = -2gs$$
$$v^2 = -2 \times 9.81 \times (-150) \text{ m/s}$$
$$v = \sqrt{2943} \qquad = \underline{54.25 \text{ m/s}}$$

$$\text{Input power} = \frac{\text{output}}{\text{efficiency}}$$
$$= 50/0.75 \qquad = \underline{66.67 \text{ MW (MJ/s)}}$$

$$66.67 \times 10^6 \text{ (J/s)} = \frac{1}{2} Mv^2$$

$$\text{Mass of water flowing/s} = \frac{2 \times 66.67 \times 10^6}{v^2} \text{ kg/s}$$

$$= \frac{2 \times 66.67 \times 10^6}{(54.25)^2}$$

$$= \frac{133.34 \times 10^6}{2943} \qquad = \underline{0.0453 \times 10^6 \text{ kg/s}}$$

$$\text{Water flowing in gal/s} = \frac{0.0453 \times 10^6}{4.54} \text{ gal/s}$$

$$\text{Rate of flow in gal/min} = \frac{0.0453 \times 10^6}{4.54} \times 60 \ = \ \underline{\textbf{0.6} \times 10^6 \textbf{ gal/min}}$$

Example 1.14

*A piece of copper having a volume of 10 cm³ was first drawn into a wire
100 m long and afterward rolled into a square copper sheet of 10 cm side.
Calculate the resistance of the wire and the resistance between opposite
faces of the sheet if the specific resistance of the copper is
1.7 x 10⁻⁶ Ω-cm.*

(contd)

Solution

<u>Wire 100 m long</u>

Resistance, $R = \rho\, l/a$

Length, $l = 100$ m

Volume $=$ area x length

10 cm$^3 = a$ x 100 x 100 cm

$\therefore$ Area, $a = \dfrac{10 \text{ cm}^3}{100 \times 100 \text{ cm}} = 1.0 \times 10^{-3}$ m^2

and $R = 1.7 \times 10^{-6} \times \dfrac{10 \times 10^3}{1.0 \times 10^{-3}} = \mathbf{1.7\ \Omega}$

<u>Copper sheet of 10 cm side</u>

Thickness of sheet $= \dfrac{\text{volume}}{\text{area}}$

$= \dfrac{10 \text{ cm}^3}{10 \times 10 \text{ cm}^2} = \underline{0.1 \text{ cm}}$

The thickness of the sheet becomes the length between opposite faces.

That is, length $= 0.1$ cm

Area $= 100$ cm^2

$\therefore \quad R = \rho\, l/a = 1.7 \times 10^{-6} \times \dfrac{0.1}{100} = \mathbf{1.7 \times 10^{-9}\ \Omega}$

$= \mathbf{0.0017\ \mu\Omega}$

Example 1.15

Two lengths of conductors, copper and aluminum, are connected in parallel to share a current of 2 A. The aluminum wire is 10 m long and 2 mm in diameter while the copper wire is 6 m long and of unknown diameter. If the current through the aluminum wire is 1.25 A, calculate the unknown diameter of the copper wire.

$[\rho = 1.6 \times 10^{-6}\ \Omega\text{-cm}$ for copper and $2.6 \times 10^{-6}\ \Omega\text{-cm}$ for aluminum]

Solution

Current in copper wire $= 2 - 1.25 = \underline{0.75 \text{ A}}$

Ratio of current in aluminum to current in copper is 1.25 : 0.75

Current is inversely proportional to resistance. That is

$I \propto {}^1/R$

$\therefore$ Resistance of copper wire $= \dfrac{1.25}{0.75}$ x resistance of aluminum wire

(contd)

Let R_1 = resistance of aluminum

and R_2 = resistance of copper

$$R_1 = \rho_1 \frac{l_1}{a_1}$$

$$R_2 = \rho_2 \frac{l_2}{a_2}$$

$$\frac{R_2}{R_1} = \frac{\rho_2 \frac{l_2}{a_2}}{\rho_1 \frac{l_1}{a_1}}$$

$$= \frac{\rho_2\, l_2\, a_1}{\rho_1\, l_1\, a_2}$$

$$\frac{1.25}{0.75} = \frac{1.6 \times 10^{-6} \times 600 \times a_1}{2.6 \times 10^{-6} \times 10^3 \times a_2}$$

$$\frac{1.25}{0.75} = \frac{1.6 \times 6\, a_1}{26\, a_2}$$

$$= \frac{9.6\, a_1}{26\, a_2}$$

$$\therefore \quad a_1/a_2 = \frac{1.25 \times 26}{0.75 \times 9.6} \quad \text{cm}^2$$

$$= \frac{32.5}{7.2} \qquad = \underline{4.514\ \text{cm}^2}$$

$$\text{Area, } a = \frac{\pi d^2}{4}$$

Hence diameter, $d \propto \sqrt{a}$

$$\therefore \quad a_1/a_2 = \frac{4.514}{1}$$

and $a_2 = a_1 \times {}^1/4.514$

$$\therefore \quad d_2 = d_1 \times \sqrt{({}^1/4.514)}$$

$$= 0.471 \times 2\ \text{mm} \qquad = \underline{\textbf{0.942 mm}}$$

Diameter of copper wire is 0.942 mm

Example 1.16

A conductor rail has a cross-section of 15 cm² and a specific resistance of 7.6 x 10⁻⁶ Ω-cm at 0 °C. If the temperature coefficient of the material is 0.005/ °C, estimate its resistance in ohms per kilometer (Ω/km) when the temperature is 50 °C.

(contd)

Solution

Given the value of ρ at $0^{\circ}C$, the specific resistance of the conductor at $t^{\circ}C$ is obtained from the expression

$$\rho_t = \rho_0 (1 + \alpha t) \quad (\alpha = \text{temperature coefficient at } 0^{\circ}C)$$

$$\therefore \quad \rho_{50} = 7.6 \times 10^{-6} (1 + 0.005 \times 50)$$

$$= 7.6 \times 10^{-6} (1 + 0.25)$$

$$= 7.6 \times 1.25 \times 10^{-6} \qquad = \underline{9.5 \times 10^{-6} \ \Omega\text{-cm}}$$

Resistance of 1 km length at $50^{\circ}C$ is given by the equation

$$R = \rho\, l / a$$

$$= \frac{9.5 \times 10^{-6} \times 1000 \times 10^2}{15} \qquad = \underline{\textbf{0.063 } \Omega}$$

<u>*Alternatively*</u>

Since the length is in kilometers, convert all units to meters.

Thus $\qquad \rho_{50} = 9.5 \times 10^{-6} \ \Omega\text{-cm} \quad = 9.5 \times 10^{-8} \ \Omega\text{-m}$

$\qquad$ Area, $a = 15 \ \text{cm}^2 \qquad = 15 \times 10^{-4} \ \text{m}^2$

$\qquad$ Length, $l = 10^3 \ \text{m}$

Resistance of 1 km length at $50^{\circ}C$ is

$$R = \frac{9.5 \times 10^{-8} \times 10^3}{15 \times 10^{-4}}$$

$$= 0.63 \times 10^{-1} \qquad = \underline{\textbf{0.063 } \Omega}$$

Example 1.17

An electric kettle holds 2 lb of water and is connected to a 240 V mains. It is required to raise the temperature of the water to the boiling point (212 °F) in 10 min; the ambient temperature is taken as 60 °F. Assuming an efficiency of 90%, calculate the resistance of the heating element of the kettle.

Solution

$$\text{Heat given to water} = \text{weight (lb)} \times \text{temperature rise } (^{\circ}F)$$

$$= 2 \times [(212 - 60)^{\circ}F] \ \text{BTU}$$

$$= 2 \times 152 \qquad \qquad = \underline{304 \ \text{BTU}}$$

$$\text{Heat in calories} = \text{BTU} \times 252$$

$$= 304 \times 252 \qquad \qquad = \underline{76608 \ \text{cal}}$$

$$\text{Input heat} = \frac{\text{output heat}}{\text{efficiency}} = \frac{76608}{0.9} \qquad = \underline{85120 \ \text{cal}}$$

19

(contd)

$$\text{Energy in joules} = \text{calories} \times 4.18 \qquad (1 \text{ cal} = 4.18 \text{ J})$$
$$= 85120 \times 4.18 \qquad = \underline{355801.6 \text{ J}}$$

$$\text{Energy in watts} = \text{joules/sec}$$
$$= \frac{355801.6}{10 \text{ (min)} \times 60} \text{ watts (J/s)}$$
$$= \frac{355801.6}{600} \qquad = \underline{593 \text{ W}}$$

$$\text{Power, } P = V^2/R$$
$$\therefore \text{Resistance, } R = V^2/P$$
$$\text{R of heating element} = \frac{240^2}{593} \qquad = \underline{\mathbf{97.13 \ \Omega}}$$

Example 1.18

An immersion heater rated at 5 kW takes 2 h to raise 15 gal of water from 10 °C to 90 °C. Calculate the efficiency of the heater

    (a)   using SI (MKS -Meter, Kilogram, Second) units

    (b)   using Imperial (FPS -Foot, Pound, Second) units

[Take 1 gal of water = 10 lb; 1 kcal = 4187 J]

Solution

(a)   <u>System International units (SI units)</u> - MKS

$$\text{Heat energy given to water} = \text{weight of water} \times \text{temperature rise}$$
$$= 15 \times 10 \text{ (lb)} \times 453.6 \text{ (g)} \times (90 - 10)^{\circ}\text{C}$$
$$= \frac{150 \times 453.6 \times 80}{10^3} \text{ kcal}$$
$$= \frac{68040 \times 80}{10^3} \qquad = \underline{5443.2 \text{ kcal}}$$

$$\text{Electrical input to heater} = \text{power (watts)} \times \text{time (sec)}$$
$$= 5 \times 10^3 \text{ (J/sec)} \times 2 \times 3600 \text{ (sec)}$$
$$= 5000 \times 2 \times 3600 \qquad = \underline{36 \times 10^6 \text{ J}}$$
$$= \frac{36 \times 10^6}{4187} \qquad = \underline{8598.04 \text{ kcal}}$$

$$\text{Efficiency} = \frac{\text{output}}{\text{input}}$$
$$= \frac{5443.2 \text{ (kcal)}}{8598.04 \text{ (kcal)}} \times 100\% \qquad = \underline{\mathbf{63.31\%}}$$

(contd)

(b)   <u>Imperial units</u> - FPS

$$\text{Temperature rise} = (90 - 10)^{\circ}C = \underline{80^{\circ}C}$$

Equivalent temperature rise in $^{\circ}F = {}^{9}/5 \times 80^{\circ}C = \underline{144^{\circ}F}$

Note: The conversion from $^{\circ}C$ to $^{\circ}F$ is as follows:

$$^{\circ}F = (^{\circ}C \times {}^{9}/5) + 32$$

Equivalent temperature rise $= [(90 \times {}^{9}/5) + 32] - [(10 \times {}^{9}/5) + 32]\ ^{\circ}F$

$$= (194 - 50)\ ^{\circ}F = \underline{144^{\circ}F}$$

$$\text{Heat energy given to water} = 15\ (gal) \times 10\ (lb) \times 144^{\circ}F$$
$$= \underline{21600\ BTU}$$

$$\text{Electrical energy input to heater} = 5\ (kW) \times 2\ (h) = \underline{10\ kWh}$$

Now

$$1\ kWh = 3.6 \times 10^{6}\ J$$
$$1\ BTU = 252\ cal$$
$$1\ cal = 4187 \times 10^{-3}\ J$$

$$\therefore \quad kWh = \frac{3.6 \times 10^{6}}{BTU}$$

$$= \frac{3.6 \times 10^{6}}{252 \times 4.187} = \underline{3411.92\ BTU}$$

$$\text{Electrical energy input to heater} = 10\ (kWh) \times 3411.92$$
$$= \underline{34119.2\ BTU}$$

$$\text{Efficiency} = \frac{\text{output}}{\text{input}}$$

$$= \frac{21600}{34119.2} \times 100\% = \underline{\mathbf{63.31\%}}$$

Example 1.19

An immersion heater is placed in a calorimeter containing 146 g of water at 20 °C. A constant current was passed through the heater and measurements were noted as follows:

> *Current*   *= 2.15 A*
>
> *p.d between heater terminals = 5.0 V*
>
> *Heating time was*   *= 5.5 min*
>
> *Final temperature of water   = 25.5 °C*

Assuming no loss of heat, calculate the experimental value of the number of joules which are equivalent to 1 cal. [Take the water equivalent of the calorimeter to be 5 g]

21

(contd)

Solution

Total weight of water $\quad$ = actual water + water equivalent of calorimeter

$$= \; 146 + 5 \qquad = \underline{151\ g}$$

Heat produced in calorimeter = total weight x temperature rise

$$= 151 \times (25.5 - 20)^\circ C$$

$$= 151 \times 5.5 \qquad = \underline{830.5\ cal}$$

Power input to heater = p.d across coil x current

$$= 5.0 \times 2.15 \qquad = \underline{10.75\ W\ (J/sec)}$$

Heat consumption in 5.5 min = power input (J/sec) x time (sec)

$$= 10.75 \times (5.5 \times 60)$$

$$= 10.75 \times 330 \qquad = \underline{3547.5\ J}$$

Heat produced in calorimeter = heat consumption in 5.5 min

That is, $\qquad$ 830.5 cal $\; = 3547.5$ J

$$\therefore \qquad 1\ cal \; = \frac{3547.5\ J}{830.5} \qquad = \underline{\mathbf{4.272\ J}}$$

Example 1.20

An immersion heater in a residential unit rated at 5 kW is fitted in a lagged tank (that is, an externally insulated tank to minimize heat loss) and the temperature is controlled by a thermostat. The temperature of the inlet water to the tank is 12 °C and the thermostat switches the heater off at 72 °C. During a day of 24 h, 60 gal of water at 72 °C is drawn from the tank. The heat losses provide the means for the thermostat to switch the heater on for 1 h in each day. Calculate the overall efficiency and the cost of running the heater for a month of 31 days at 16 cents/kWh.

Solution

Weight of 1 gal of water $\; = 10$ lb

$$1\ lb \; = 453.6\ g$$

Temperature rise $\; = (72 - 12)^\circ C = \; 60^\circ C$

Energy required $\; = 60\ (gal) \times 10\ (lb) \times 453.6\ (g) \times 60^\circ C$

$$= 600 \times 453.6 \times 60\ cal$$

$$= 600 \times 453.6 \times 60 \times 4.18 \quad = \underline{68.26 \times 10^6\ J}$$

(contd)

$$\therefore \quad \text{Energy in kWh} \quad = \frac{68.26 \times 10^6}{3.6 \times 10^6} \qquad = \underline{18.96 \text{ kWh}}$$

$$\text{Energy loss in 24 h} \quad = 5 \text{ (kW)} \times 1 \text{ (h)} \qquad = \underline{5 \text{ kWh}}$$

$$\text{Energy input} \quad = 18.96 + 5 \qquad = \underline{23.96 \text{ kWh}}$$

$$\text{Efficiency} \quad = 1 - \frac{\text{losses}}{\text{input}}$$

$$= (1 - {}^5/23.96) \times 100\% \qquad = \underline{\mathbf{79.13\%}}$$

$$\text{Cost of energy for 31 days} \quad = \text{input} \times \text{days} \times \text{unit cost}$$

$$= 23.96 \times 31 \times 0.16 \qquad = \underline{\mathbf{\$\,118.84}}$$

Example 1.21

A pump driven by an electric motor lifts 500 gal of water per minute to a height of 165 ft. The efficiency of the pump is 90% and that of the motor 85%. The motor is supplied at 250 V. Determine (a) the output horsepower, and (b) the current taken by the motor

Solution

(a)

$$\text{Work done by pump} \quad = \text{mass} \times \text{distance lifted (height)}$$

$$= 500 \text{ (gal)} \times 10 \text{ (lb)} \times 165 \text{ ft}$$

$$= 5000 \times 165 \text{ ft-lbf}$$

$$\text{Rate of doing work} \quad = \frac{5000 \times 165}{1 \text{ (min)}} \qquad = \underline{5000 \times 165 \text{ ft-lb/min}}$$

$$1 \text{ h.p} \quad = 33000 \text{ ft-lb/min}$$

$$\text{Pump output h.p} \quad = \frac{5000 \times 165}{33000} \qquad = \underline{\mathbf{25 \text{ h.p}}}$$

(b)

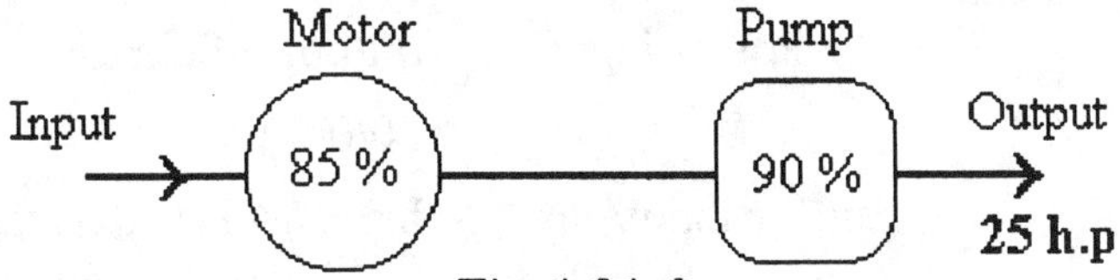

Motor   Pump

Input   85 %   90 %   Output   25 h.p

Fig. 1.21-0

<u>Refer to Fig. 1.21-0</u>

$$\text{Input to pump} \quad = \frac{\text{output}}{\text{efficiency}} = \frac{25}{0.9} \qquad = \underline{27.8 \text{ h.p}}$$

$$\text{Input hp to motor} \quad = \frac{\text{output}}{\text{efficiency}} = \frac{27.8}{0.85} \qquad = \underline{32.71 \text{ h.p}}$$

$$\text{Current, } I \quad = \frac{\text{power}}{\text{voltage}} = \frac{32.71 \times 746}{250} = \underline{\mathbf{97.6 \text{ A}}}$$

Example 1.22

A diesel engine which drives a generator has an output of 50 h.p
operating at an efficiency of 32%. The generator has an efficiency of
82% and supplies a load circuit at 240 V.
The diesel engine uses light fuel oil (LFO) with a calorific value of
19600 BTU/lb. Calculate

    (a)   the output of the generator in kilowatts

    (b)   the current in the generator load circuit

    (c)   the fuel consumption in gallons per day

(Weight of LFO = 8.3 lb/gal; 1 BTU = 1055 J)

Solution

(a)   <u>Generator</u>

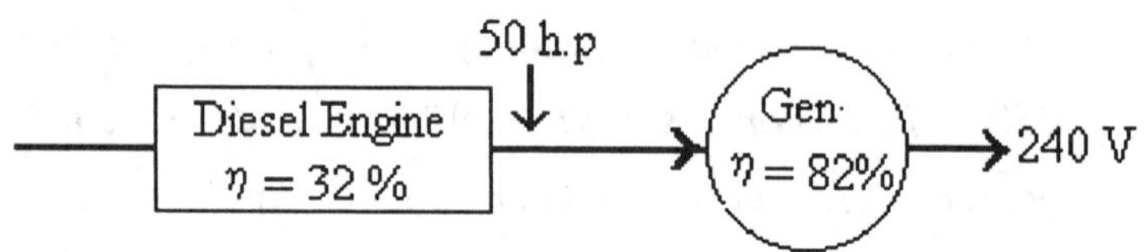

Fig. 1.22-0. Diesel generator

<u>Refer to Fig. 1.22-0</u>

Input to generator   = 50 h.p

Output of generator   = (input x efficiency) x 0.746 kW

    = 50 x 0.82 x 0.746 = **30.586 kW**

(b) Current in generator load circuit = $\dfrac{\text{power}}{\text{voltage}}$

    = $\dfrac{30.586 \times 10^3}{240}$ = **127.4 A**

(c)   <u>Diesel Engine</u>

Power input   = output/efficiency

    = 50/0.32   = 156.25 h.p

Energy input/day   = h.p x 746 (watts) x (h/day x sec)

    = 156.25 x 746 x 24 x 3600 J/day

    = 100.71×10^8 J/day

Energy input (BTU/day) = $\dfrac{100.71 \times 10^8}{1055}$ = 9.54597×10^6

Calorific value of LFO = 19600 BTU/lb

Fuel consumption in lb/day = $\dfrac{9.54597 \times 10^6}{19600}$ = 4.87039×10^2

(contd)

$$\text{Fuel consumption in gal/day} = \frac{4.87039 \times 10^2}{8.3} = \underline{58.679 \text{ gal/day}}$$

$$\text{Fuel consumption in gal/h} = \frac{58.679}{24} = \mathbf{2.44 \text{ gal/h}}$$

Alternatively

If you are confident and have no objection in being repetitious, then the final calculation may be performed at the end of the final statement, which will give the same result. Thus

$$\text{Power input} = {}^{50}/0.32 = \underline{156.25 \text{ h.p}}$$

$$\text{Energy input/day} = 156.25 \times 746 \times 24 \times 3600 \quad \text{J/day}$$

$$\text{Energy input (BTU/day)} = \frac{156.25 \times 746 \times 24 \times 3600}{1055} \quad \text{BTU/day}$$

$$\text{Energy input in lb/day} = \frac{156.25 \times 746 \times 24 \times 3600}{1055 \times 19600} \quad \text{lb/day}$$

$$\text{Energy input in gal/day} = \frac{156.25 \times 746 \times 24 \times 3600}{1055 \times 19600 \times 8.3} \quad \text{gal/day}$$

$$\text{Energy input in gal/h} = \frac{156.25 \times 746 \times 24 \times 3600}{1055 \times 19600 \times 8.3 \times 24} = \mathbf{2.44 \text{ gal/h}}$$

Example 1.23

A diesel engine with an overall efficiency of 25% uses 20 lb of fuel oil in 1 h to drive a generator. If the calorific value of the fuel is 16500 BTU/lb, determine the output horsepower developed by the engine. [1 BTU = 778 ft-lbf]

Solution

$$\text{Heat energy of fuel} = 20 \text{ (lb)} \times 16500 \text{ (BTU/lb)}$$
$$= 330 \times 10^3 \text{ BTU}$$
$$= 330 \times 10^3 \times 778 \text{ ft-lbf}$$
$$= \frac{330 \times 10^3 \times 778}{60} \text{ ft-lbf/min}$$

$$1 \text{ Horsepower} = 33000 \text{ ft-lbf/min}$$

$$\therefore \quad \text{Horsepower} = \frac{330 \times 10^3 \times 778}{60 \times 33000} = \underline{129.67 \text{ h.p}}$$

$$\text{Output hp of engine} = \text{input} \times \text{efficiency}$$
$$= 129.67 \times \frac{25}{100} = \mathbf{32.42 \text{ h.p}}$$

Example 1.24

Two coils connected in series have resistances of 400 Ω and 200 Ω and temperature coefficients of 0.1% and 0.4%, respectively, at 15 °C. Calculate (a) the resistance of the combination at 65 °C and (b) the effective temperature coefficient of the combination.

Solution

$$\text{Let } R_2 \quad = \text{ resistance of a coil at 65°C } (t_2)$$
$$R_1 \quad = \text{ resistance of a coil at 15°C } (t_1)$$
$$\alpha \quad = \text{ temperature coefficient of the coil at 15°C } (t_1)$$

From Eq. 1.0-4

$$R_{t2} \quad = R_{t1} [1 + \alpha (t_2 - t_1)]$$

Resistance R_2 of 400 Ω coil at 65°C is

$$R_2 \quad = R_1 [1 + \alpha (t_2 - t_1)]$$
$$= 400 [1 + 0.1/100 \ (65 - 15)°C]$$
$$= 400 [1 + 0.001 \ (50)] \qquad = \underline{420 \ \Omega}$$

Similarly, resistance R_2 of 200 Ω coil at 65°C is

$$R_2 \quad = 200 [1 + 0.004 \ (50)]$$
$$= 200 \times 1.2 \qquad\qquad = \underline{240 \ \Omega}$$

(a) Therefore, resistance of combination at 65°C

$$= (420 + 240) \ \Omega \qquad\qquad = \underline{\textbf{660 } \Omega}$$

(b) <u>Temperature coefficient (α) for the combination</u>

The combined resistance R_1 at 15°C is

$$= (400 + 200) \ \Omega \qquad\qquad = \underline{600 \ \Omega}$$

The combined resistance R_2 at 65°C is

$$= (420 + 240) \ \Omega \qquad\qquad = \underline{660 \ \Omega}$$

$$\text{Now} \qquad R_2/R_1 \ = 1 + \alpha (t_2 - t_1)$$
$$660/600 \quad = 1 + \alpha (65 - 15)$$
$$\therefore \qquad 50 \alpha \quad = 660/600 - 1$$
$$\text{and} \qquad\qquad \alpha \quad = \frac{1.1 - 1}{50} \qquad = \underline{\textbf{0.002}} \quad \text{or} \quad \underline{\textbf{0.2\%}}$$

Example 1.25

The field coil of a motor has a resistance of 10 Ω at 15 °C. How much will the resistance increase if the motor attains an average temperature of 60 °C when running? [Take α = 0.00427/ °C referred to 0 °C]

Solution

Let $\quad R_0$ = resistance of coil at 0°C

$\quad R_{15}$ = resistance of coil at 15°C

$\quad R_{60}$ = resistance of coil at 60°C

$\quad$ t = temperature at t°C (15°C and 60°C)

Then $\quad R_{15} = R_0 (1 + \alpha t)$

$\quad R_{60} = R_0 (1 + \alpha t)$

$$R_{15}/R_{60} = \frac{R_0 (1 + 15\alpha)}{R_0 (1 + 60\alpha)} \qquad \text{(refer to Eq. 1.0-3)}$$

$$= \frac{1 + 15\alpha}{1 + 60\alpha}$$

$$= \frac{1/\alpha + 15}{1/\alpha + 60}$$

Now $\quad 1/\alpha = 1/0.00427 \qquad = \underline{234.2}$

$\therefore \quad R_{15}/R_{60} = \dfrac{234.2 + 15}{234.2 + 60} \qquad = \dfrac{\underline{249.2}}{294.2}$

and $\quad R_{60} = R_{15} \times \dfrac{294.2}{249.2}$

$$= 10 \times 1.181 \qquad = \underline{\underline{11.81\ \Omega}}$$

Example 1.26

The field coil of a 240 V shunt motor takes 1.6 A when first switched on, the ambient temperature being 20 °C. After several hours, the field current was found to remain steady at 1.25 A. Determine the temperature of the winding.
[Assume temperature coefficient of copper = 0.004/ °C at 0 °C]

Solution

Coil resistance at 20°C $\quad = V/I = 240/1.6 \qquad = \underline{150\ \Omega}$

Resistance of field coil at a higher temperature t_2 is

$$= 240/1.25 \qquad = \underline{192\ \Omega}$$

(contd)

$$\text{Now} \qquad R_{20}/R_{t2} = \frac{R_0\,(1 + \alpha t_1)}{R_0\,(1 + \alpha t_2)}$$

$$\frac{150}{192} = \frac{1 + 0.004 \times 20}{1 + 0.004 \times t_2}$$

$$150\,(1 + 0.004t_2) = 192\,(1 + 0.004 \times 20)$$

$$150 + 0.6t_2 = 192\,(1.08)$$

$$\therefore \qquad t_2 = \frac{(192 \times 1.08) - 150}{0.6}$$

$$= \frac{207.36 - 150}{0.6}$$

$$= \frac{57.36}{0.6} \qquad = \mathbf{95.6^oC}$$

Note:

It is of interest to recognize and compare the variables and solutions in Examples 1.26 and 1.27.

Example 1.27

A coil of insulated copper wire has a resistance of 150 Ω at 20 °C. When the coil is connected across a 240 V supply, the current after several hours is 1.25 A. Calculate the average temperature throughout the coil, assuming the temperature coefficient of resistance of copper at 20 °C to be 0.0039/ °C.

Solution

$$\text{Resistance of coil } R_1 \text{ at } 20^oC = \underline{150\ \Omega}$$

$$\text{Resistance of coil } R_2 \text{ at } t_2{}^oC = V/I$$

$$= 240/1.25 \quad = \underline{192\ \Omega}$$

$$R_2 = R_1\,[\,1 + \alpha\,(t_2 - t_1)]$$

$$192 = 150\,[1 + 0.0039\,(t_2 - 20)]$$

$$= 150\,[1 + 0.0039t_2 - 0.078]$$

$$= 150 + 0.585t_2 - 11.7$$

$$= 138.3 + 0.585t_2$$

$$t_2 = \frac{192 - 138.3}{0.585} = \frac{53.7}{0.585} \quad = \mathbf{91.8^oC}$$

Therefore, the average temperature is 91.8⁰C.

Example 1.28

An aluminum conductor has a resistance of 3.6 Ω at 20 °C. Determine the resistance at 50 °C if the temperature co-efficient of resistance of aluminum is 0.00403/ °C at 20 °C.

Solution

$$\text{Resistance } R_1 \text{ at } 20°C \;=\; 3.6\ \Omega$$
$$\text{Resistance } R_2 \text{ at } 50°C \;=\; \text{unknown}$$

$$\text{Now} \quad R_2 \;=\; R_1\,[1 + \alpha\,(t_2 - t_1)]$$
$$=\; 3.6\,[1 + 0.00403\,(50 - 20)]$$
$$=\; 3.6\,[1 + 0.00403 \times 30]$$
$$=\; 3.6\,[1 + 0.1209]$$
$$=\; 3.6 \times 1.1209 \qquad =\; \underline{\mathbf{4.035\ \Omega}}$$

Example 1.29

Estimate the length and diameter of the wire needed for a 1 kW heating element for a 230 V circuit, the permissible current density being 14 A/mm² and the specific resistance 120 $\mu\Omega$-cm when hot. If the resistance (hot) is 1.3 times the resistance at 15 °C and the working temperature is 600 °C, find the average temperature coefficient.

Solution

$$\text{Power rating of element} \;=\; 1\ kW \qquad =\; \underline{1000\ W}$$
$$\text{Current through element} \;=\; P/V = {}^{1000}/230 \quad =\; \underline{4.35\ A}$$
$$\text{Cross-sectional area of wire} \;=\; {}^{current}/\text{current density}$$
$$=\; {}^{4.35}/14 \quad =\; 0.31\ mm^2\ (31 \times 10^{-4}\ cm^2)$$

$$\text{Area} \;=\; \pi d^2/4$$
$$d^2 \;=\; \frac{4 \times \text{area}}{\pi}$$

$$\therefore \quad \text{Diameter of wire, } d \;=\; \sqrt{[4 \times {}^{0.31}/3.14]}$$
$$=\; \sqrt{[0.3949]}\ mm \qquad =\; \underline{\mathbf{0.628\ mm}}$$

$$\text{Resistance of element (hot)} \;=\; V/I = {}^{230}/4.35 \quad =\; \underline{52.87\ \Omega}$$

(contd)

From Eq. 1.0-0 $\quad R = \rho\, l/a$

$\therefore$ Length of conductor, $l = R\, a/\rho$

$$= \frac{52.87 \times 31 \times 10^{-4}}{120 \times 10^{-6}}$$

$$= \frac{52.87 \times 31 \times 10^{2}}{120} = \mathbf{1365.8\ cm}$$

Let $\quad R_2 = $ resistance at working temperature, 600°C (hot)

$\quad\quad\quad R_1 = $ resistance at 15°C

Now $\quad R_2$ when hot (600°C) is 1.3 times resistance R_1 at 15°C

Therefore, the ratio of $R_2 : R_1$ is 1.3. Hence

$$R_2 = R_1 [1 + \alpha\,(t_2 - t_1)]$$

and $\quad R_2/R_1 = 1 + \alpha\,(600 - 15)$

$\therefore \quad\quad 1.3 = 1 + (600\alpha - 15\alpha)$

$$= 1 + 585\alpha$$

and $\quad\quad \alpha = \dfrac{1.3 - 1}{585} = \mathbf{0.0005128}$

The ambient temperature was taken as 15°C. Therefore the temperature coefficient of resistance is 0.0005128/°C referred at 15°C.

Example 1.30

The instrumentation room in a hospital measures 2200 cu ft and is to be maintained at a temperature of 20 °F above the temperature of the intake air. The vent system is designed to change the air in half-hour intervals. Calculate the rating of a radiator suitable for this room, neglecting radiation loss. Take density of air as 0.08 lb per cu ft and specific heat of air as 0.26 BTU/lb. [1 BTU = 252 cal; 4.2 J = 1 cal]

Solution

Weight of air to be changed every half-hour is

$\quad\quad = $ volume x density of air

$\quad\quad = 2200 \times 0.08 \quad\quad\quad\quad = \underline{176\ lb}$

BTU of heat required every half-hour is

$\quad\quad = $ weight of air x specific heat x temperature (°F)

$\quad\quad = 176\ (lb) \times 0.26\ (BTU/lb) \times 20$

$\quad\quad\quad\quad\quad\quad\quad\quad\quad\quad = \underline{915.2\ BTU}$

(contd)

Electrical energy required every half-hour is

$$= 915.2 \text{ (BTU)} \times 252 \text{ (cal)} \times 4.2 \text{ (J)}$$

Rate of energy dissipation is

$$= (\text{joules/sec}) \quad \text{watts}$$

$$= \frac{915.2 \times 252 \times 4.2}{\frac{1}{2} \text{ (h)} \times 60 \text{ (sec)}}$$

$$= \frac{915.2 \times 252 \times 4.2}{30 \times 60} \qquad = \underline{\mathbf{538.14 \text{ J/sec or } W}}$$

Therefore the rating of the radiator is 538.14 W.

Example 1.31

It is required to maintain a loading of 5 kW in a heating unit at an initial temperature of 15°C from a 220 V supply. After the unit is switched on and allowed to settle down to a steady state, it is found that a voltage of 240 V is now necessary to maintain this loading. Assuming that the temperature coefficient of the heating element is 0.0006/°C, estimate the final temperature of the heating element.

Solution

$$\text{Let} \quad R_1 = \text{resistance at } 15°C$$

$$\text{Power,} \quad P = V^2/R_1$$

$$\therefore \quad R_1 = V^2/P \quad = 220^2/5000 = \underline{9.68 \ \Omega}$$

$$\text{Let} \quad R_2 = \text{resistance at } t_2°C$$

$$\text{Then} \quad R_2 = V^2/P \quad = 240^2/5000 = \underline{11.52 \ \Omega}$$

$$\text{Now} \quad R_2/R_1 = \frac{R_0 \, (1 + \alpha t_2)}{R_0 \, (1 + \alpha t_1)}$$

$$R_2/R_1 = \frac{1 + \alpha t_2}{1 + \alpha t_1}$$

$$R_1 \, (1 + \alpha t_2) = R_2 \, (1 + \alpha t_1)$$

$$9.68 \, (1 + 0.0006 t_2) = 11.52 \, (1 + 0.0006 \times 15)$$

$$9.68 + 0.005808 t_2 = 11.52 \, (1 + 0.009)$$

$$0.005808 t_2 = 11.6237 - 9.68$$

$$t_2 = \frac{11.6237 - 9.68}{0.005808} \qquad = \underline{\mathbf{334.7°C}}$$

Therefore, the final temperature of the heating element is 334.7°C.

Example 1.32

Determine the loading necessary for an electric heater to raise the temperature of 1 gal of water from 20°C to 90°C in 10 min. Take the heat loss due to radiation during this period as 60000 J and the water equivalent of the heater as 150 g. Calculate the approximate efficiency of the operation. [1 gal of water = 10 lb; 1 lb = 453.6 g]

Solution

$$
\begin{aligned}
\text{Weight of water} &= 1\,(\text{gal}) \times 10\,(\text{lb}) \times 453.6\,(\text{g}) = \underline{4536\ g} \\
\text{Water equivalent of heater} &= 150\ g \\
\text{Total weight} &= (4536 + 150)\ g = \underline{4686\ g} \\
\text{Heat required in 10 min} &= \text{weight of water} \times \text{temperature rise} \\
&= 4686 \times (90 - 20) = \underline{(4686 \times 70)\ cal} \\
\text{Electrical energy required} &= (4686 \times 70) \times 4.18\ J \\
\text{Heat loss due to radiation} &= 60000\ J \\
\text{Total energy consumption} &= (4686 \times 70 \times 4.18) + 60000\ J \\
\text{Rate of energy dissipation} &= \frac{(4686 \times 70 \times 4.18) + 60000}{10\,(\text{min}) \times 60\,(\text{sec})} \\
&= \frac{1431123.6}{600} = \mathbf{2385.2\ W}
\end{aligned}
$$

∴ **Rating of heater is 2.385 kW**

$$
\begin{aligned}
\text{Efficiency} &= \frac{\text{output}}{\text{input}} = \frac{\text{heat actually given to water}}{\text{total energy consumption}} \\
&= \frac{(4536 \times 70) \times 4.18\ J}{1431123.6\ J} \times 100\% \\
&= \underline{\mathbf{92.74\%}}
\end{aligned}
$$

Example 1.33

A coil with a mean diameter of 25.46 cm consists of 400 turns of wire of cross-sectional area 0.006 cm² and a specific resistance of 1.6 x 10⁻⁶ Ω-cm. If the specific heat of the wire is 0.09 cal/g and its density is 8.9 g/cm³, determine the rate of temperature rise in °C per minute if the pd across the terminals of the coil is 20.5 V. Assume that the resistance of the coil remains constant during the rise in temperature and neglect the weight of the insulating material and heat loss due to radiation and conduction. [Take 1cal = 4.18 J]

(contd)
Solution

$$\text{Resistance, R} = \rho\, {}^{l}/a$$

$$\text{Circumference of coil} = \pi d = 3.142 \times 25.46 = \underline{80\ cm}$$

$$\text{Length of wire} = \text{circumference} \times \text{turns}$$

$$= 80 \times 400 = \underline{32 \times 10^3\ cm}$$

$$\therefore \quad R = \frac{1.6 \times 10^{-6} \times 32 \times 10^3}{0.006}$$

$$= \frac{1.6 \times 32}{6} = \underline{8.533\ \Omega}$$

$$\text{Current in coil} = {}^{V}/R = {}^{20.5}/8.533 = \underline{2.4\ A}$$

$$\text{Energy consumed/min} = V \times I \times \text{time (sec)}$$

$$= 20.5 \times 2.4 \times 60\ J$$

$$\text{Calories produced/min} = \frac{20.5 \times 2.4 \times 60}{4.18} = \underline{706.22\ cal/min}$$

$$\text{Volume of wire} = \text{length} \times \text{area}$$

$$= 32 \times 10^3 \times 0.006 = \underline{192\ cm^3}$$

$$\text{Mass (weight) of wire} = \text{volume} \times \text{density}$$

$$= 192 \times 8.9 = \underline{1708.8\ g}$$

$$\text{Heat dissipated in wire/min} = \text{calories produced/min}$$

$$\therefore \quad \text{Mass} \times \text{specific heat} \times \text{temperature rise/min}$$

$$= 706.22\ cal/min$$

$$1708.8 \times 0.09 \times \text{temperature (t) rise/min}$$

$$= 706.22$$

$$\text{Temperature (t) rise/min} = \frac{706.22}{1708.8 \times 0.09}$$

$$= \frac{706.22}{153.792} = \underline{\mathbf{4.592^{o}C/min}}$$

Example 1.34

The resistance of a coil is increased by 20% when a current is passed through it for 16 sec. Assume that the average resistance is 1.5 Ω and the current is constant. Calculate the current in the coil if the mass of the coil is 1 kg. The specific heat of the material is 0.15 cal/g and the temperature coefficient is 0.5% per ^{o}C. [1 cal = 4.18 J]

(contd)

Solution

$$\text{Let} \quad R_1 = \text{initial resistance at } t_1 {}^\circ C$$
$$\text{and} \quad R_2 = \text{final resistance at } t_2 {}^\circ C$$
$$\text{Then} \quad R_2 = R_1[1 + \alpha (t_2 - t_1)]$$
$$R_2/R_1 = 1 + (0.5/100)(t_2 - t_1)$$
$$= 1 + 0.005 (t_2 - t_1)$$

Assume the initial resistance R_1 is unity (1).

Then the final resistance R_2, having increased by 20%, is

$$R_2 = R_1 + 20\%$$
$$= 1 + 20/100 = 1 + 0.2 \qquad = \underline{1.2 \ \Omega}$$
$$\text{Thus} \quad R_2/R_1 = 1.2/1$$
$$\therefore \quad 1.2 = 1 + 0.005 (t_2 - t_1)$$
$$t_2 - t_1 = \frac{1.2 - 1}{0.005} \qquad = \underline{40 {}^\circ C}$$

$$\text{Rise of temperature in coil} = t_2 - t_1 \qquad = \underline{40 {}^\circ C}$$
$$\text{Heat dissipation in coil} = \text{mass x specific heat x temperature rise}$$
$$= 1000 \times 0.15 \times 40 \qquad = \underline{6000 \ cal}$$
$$\text{Electrical energy dissipated} = 6000 \times 4.18 \qquad = \underline{25080 \ J}$$
$$\text{Power consumed} = (\text{joules}/\text{sec}) \ \text{watts}$$
$$= 25080/16 \qquad = \underline{1567.5 \ W}$$
$$\text{Now,} \quad \text{Power} = I^2 R$$
$$I^2 R = 1567.5$$
$$I^2 = 1567.5/R$$
$$= 1567.5/1.5 \qquad = \underline{1045}$$
$$\therefore \quad I = \sqrt{1045} \qquad = \underline{\mathbf{32.326 \ A}}$$

$$\underline{\textbf{Hence, current in coil is 32.326 A}}$$

Example 1.35

The electric furnace of a recycling plant smelts 100 kg of tin per hour from an initial temperature of 20 °C. Calculate the power required for this smelt and the cost of recycling tin at 20 cents per kilowatt-hour.

Specific heat of tin = 0.056 cal/g

Melting point of tin = 240 °C

Latent heat of fusion = 14.5 kcal/kg

Thermal efficiency = 72%

(contd)

Solution

Heat required to raise 100 kg of tin from 20°C to melting point

$\quad$ = mass x specific heat x temperature rise

$\quad$ = $100 \times 10^3 \times 0.056 \times (240 - 20)$

$\quad$ = $100 \times 10^3 \times 0.056 \times 220$

$\qquad\qquad$ = $\underline{1232 \times 10^3 \text{ cal}}$

Heat to smelt 100 kg of tin $\quad$ = mass x latent heat of fusion

$\quad$ = $100 \text{ (kg)} \times 14.5 \times 10^3 \text{ (cal/kg)}$

$\qquad\qquad$ = $\underline{1450 \times 10^3 \text{ cal}}$

Total effective heat used per hour $\quad = (1232 \times 10^3) + (1450 \times 10^3)$

$\qquad\qquad$ = $\underline{2682 \times 10^3 \text{ cal}}$

Heat generated per hour $\quad = \dfrac{\text{effective heat}}{\text{efficiency}}$

$\quad = \dfrac{2682 \times 10^3}{0.72}$ $\qquad$ = $\underline{3725 \times 10^3 \text{ cal}}$

Energy consumed per hour = $3725 \times 10^3 \times 4.18$ J

$\qquad\qquad$ = $\underline{15.571 \times 10^6 \text{ J}}$

Rate of energy dissipation $\quad$ = joules/sec (watts)

$\quad = \dfrac{15.571 \times 10^6}{3600}$ $\qquad$ = $\underline{4.325 \times 10^3 \text{ W}}$

Energy used to smelt 100 kg per hour is

$\quad$ = 4.325 kWh

Cost of smelt $\quad$ = 4.325×20 $\qquad$ = **<u>86.5 cents</u>**

PROBLEMS   1

1. An express train passes station A with a velocity of 20 m/s and station B 1 km away with a velocity of 30 m/s, the acceleration being uniform. Determine (a) the average velocity of the train, (b) the time taken for the train to travel the distance between the two stations, and (c) the acceleration of the train.

   [Ans. (a) 25 m/s, (b) 40 s, (c) 0.25 m/s²]

2. A train takes 5 min to complete the journey between two stations 3 km apart. For the first 30 s the train moves with constant acceleration, while a uniform retardation in the last 20 s brings the train to a standstill. For the remaining portion of the journey the train maintains a uniform speed. Calculate the value of (i) the uniform speed, (ii) the acceleration, and (iii) the retardation. (iv) Illustrate your answer with a speed-time graph and use this graph to obtain the distance traveled in the first and last minutes of the train's motion.

   [Ans. (i) 10.9 m/s, (ii) 0.363 m/s², (iii) 0.545 m/s², (iv) 490 m, 545 m]

3. In a test to determine Joule's Equivalent, a current of 10 A flowing in a resistance of 2.5 Ω was found to raise the temperature of 500 cm³ of water 35°C in 5 min. Calculate the value of Joule's Equivalent from these data. Assume that 1 cm³ of water weighs 1 g.

   [Ans. 4.286 J/cal]

4. The resistance of the heater element of a 240 V kettle is 72 Ω. Taking the efficiency as 90%, determine the time taken to heat 2.5 lb of water from 60°F to the boiling point.

   [Ans. 9.31 min or 9 min 18.6 sec]

5. An electric soldering iron is heated on a 240 V mains and takes a current of 0.5 A. The weight of the copper bit is 130 g and 40% of the heat generated is lost through conduction, convection, and radiation. If the melting point of solder is 300°C and the ambient temperature is 15°C, calculate the waiting time before the iron is ready for use. Assume the specific heat of copper is 390 J/kg.K.

   [Ans. 3.345 min or 3 min 20.7 sec]

6. An electric kettle is required to raise the temperature of 20 L of water from 15°C to the boiling point (100°C). Calculate (a) the electrical energy consumed in megajoules and in kilowatt-hours and (b) the cost of energy consumed if the charge is 35 cents/kWh. Assume the specific heat of water to be 4190 J/kg.K, 1 L of water to have a mass of 1 kg, and the efficiency of the heater to be 90%.

   [Ans. (a) 7.123 MJ, 2.198 kWh, (b) 77 cents]

7.   A pump driven by an electric motor lifts 1.5 kg of water per minute to a height of 40 m. The pump has an efficiency of 90% and the motor an efficiency of 85%. Calculate (i) the input power to the motor from a 480 V supply, and (ii) the electrical energy consumed by the motor after 8 h of running on this load.

[Ans. (i) 12.82 kW, (ii) 102.56 kWh]

8.   A contractor employed a hoist to deliver mixed cement after raising it vertically 130 ft from the ground. The efficiency of the driving motor is 85% and that of the hoist is 80%. The motor is supplied at 400 V and takes a current of 75 A. Calculate the weight of cement delivered if the motor runs for 2 h.

[Ans. 403 tons]

9.   A steam power station uses coal of a calorific value of 8.6×10^6 BTU/lb and has an overall efficiency of 2%. Calculate the daily coal consumption in tons when the average load on the station is 100 MW.
(Take 1 ton = 2240 lb; 1 BTU = 1055 J)

[Ans. 1.7 tons]

10.   The filament of a 60 W lamp has a normal working temperature of 2000°C. Take the temperature coefficient of the material to be 0.005/°C. Find the approximate current which flows at the instant of switching on the supply to the cold lamp. Assume the temperature of the cold lamp to be 15°C.

[Ans. 2.73 A]

11.   The field winding of a 200 V shunt motor takes 2 A when first switched on, the ambient temperature being 16°C. Later the field current was found to remain steady at 1.7 A. Calculate the temperature of the field winding.

[Ans. 62.75°C]

12.   A coil is wound with 5000 ft of wire of cross-sectional area 0.05 in² and resistivity 0.67 μΩ-in. Calculate the voltage drop and power dissipated in the coil when it carries 20 A.

[Ans. 16.08 V, 321.6 W]

CHAPTER   2

DIRECT CURRENT CIRCUIT ANALYSIS

Introduction

Ohm's Law will suffice for solving simple d.c circuits. However, as the circuit becomes complex, Kirchhoff's Current and Voltage Laws are used to find a solution, more often for the current in a branch of the circuit or voltage across two nodes. Unfortunately, as the complexity of the circuit or network increases, the solution by Kirchhoff's Laws may become lengthy and cumbersome with simultaneous equations. The complexity of these equations can often be reduced by the application of other theorems.

Kirchhoff's Current Law (First Law)

The sum of the currents entering a junction in an electrical circuit is equal to the algebraic sum of the currents leaving the junction; that is, the current in the junction at any instant of time is zero.

<u>Refer to Fig. 2.0-0(a)</u>

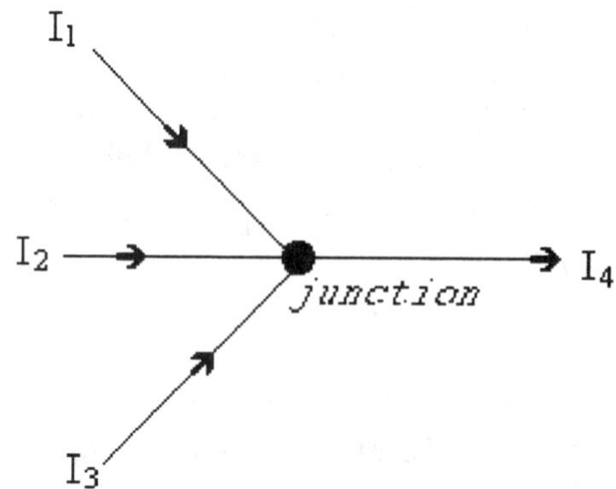

Fig. 2.0-0(a)

$$I_1 + I_2 + I_3 = I_4$$

$$I_1 + I_2 + I_3 - I_4 = 0$$

This law may be described by the following equation:

$$\Sigma I = 0$$

Kirchhoff's Voltage Law (Second Law)

In any closed loop or mesh of an electrical circuit, the algebraic sum of the IR drop (current x resistance) is equal to the resultant e.m.f in the loop.

Refer to Fig. 2.0-0(b)

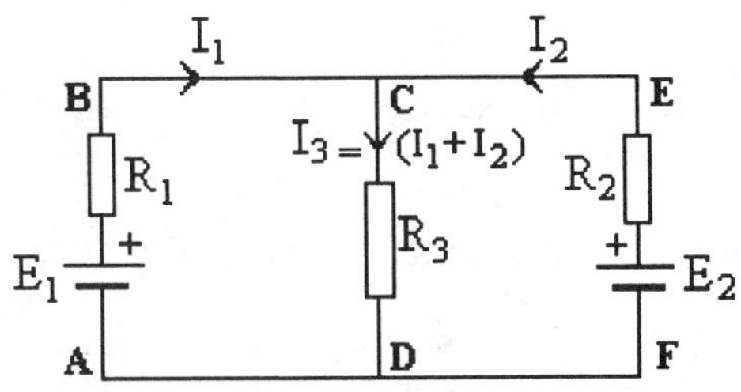

Fig. 2.0-0(b)

Consider mesh ABEFA $\qquad E_1 - E_2 = I_1 R_1 - I_2 R_2$

Consider mesh FECDF $\qquad E_2 = I_2 R_2 + I_3 R_3$

In general $\qquad \Sigma E = \Sigma IR$

Superposition Theorem

The resultant current flowing in any branch of a linear network containing more than one source of e.m.f is equal to the algebraic sum of the currents that would flow in that branch by each source acting separately with all the other sources short-circuited and replaced by their internal resistances.

Refer to Fig. 2.0-1(a), (b) and (c)

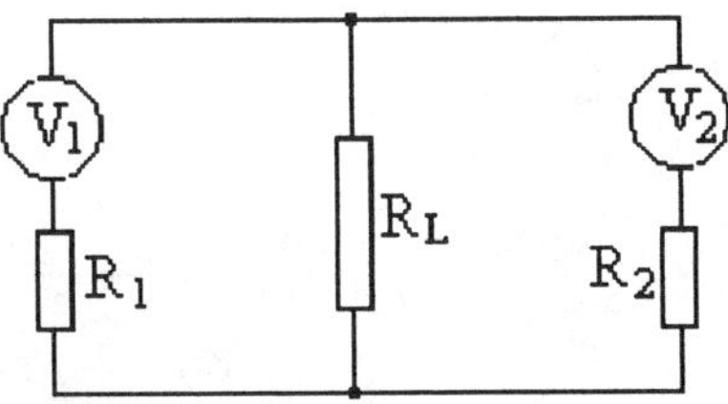

Fig. 2.0-1(a). Network with two sources of e.m.f

By applying the superposition theorem to Fig. 2.0-1(a), we obtain Fig. 2.0-1(b), in which the source V_2 is removed and replaced by its internal resistance R_2, thus allowing source V_1 to act alone.

(contd)

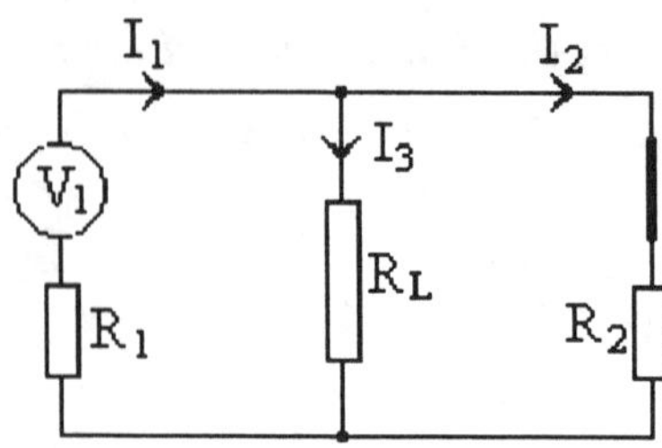

Fig. 2.0-1(b). Network with V_1 acting alone

Total resistance in the circuit $= R_1 + (R_2 R_L)/(R_2 + R_L)$

$$I_1 = \frac{V_1}{R_1 + \left(\dfrac{R_2 R_L}{R_2 + R_L}\right)}$$

$$I_2 = I_1 \times \frac{R_2}{R_2 + R_L}$$

$$I_3 = I_1 - I_2$$

Refer to Fig. 2.0-1(c)

In this circuit V_1 is now removed and V_2 is acting alone; the suggested direction of current flow in each branch is indicated by the arrowheads.

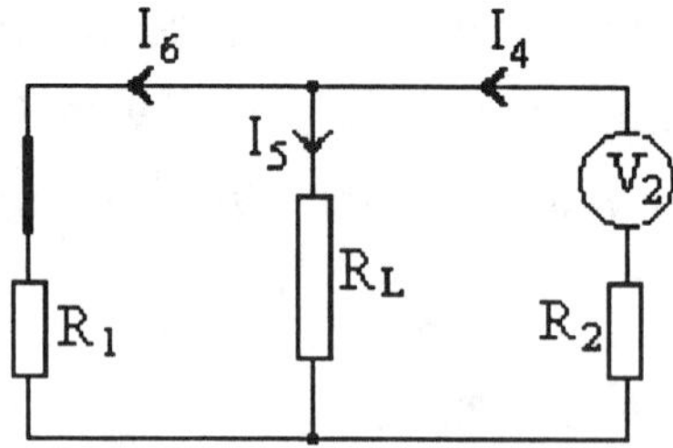

Fig. 2.0-1(c). Network with V_2 acting alone

Refer to Fig. 2.0-1(c)

$$I_4 = \frac{V_2}{R_2 + \left(\dfrac{R_1 R_L}{R_1 + R_L}\right)}$$

$$I_5 = I_4 \times \frac{R_1}{R_1 + R_L}$$

$$I_6 = I_4 - I_5$$

By superimposing Fig. 2.0-1(b) on Fig. 2.0-1(c) we obtain the resultant current in each branch of the network as shown below.

(contd)

$$\text{Resultant current through } V_1 = I_1 - I_6$$
$$\text{Resultant current through } V_2 = I_4 - I_3$$
$$\text{Resultant current in } R_L = I_2 + I_5$$

Note: If the resultant current is negative, it indicates that the
direction of current flow shown by the arrowhead is reversed.

Thevenin's Theorem

An active network having a pair of terminals may be replaced by an equivalent circuit having a constant-voltage generator of e.m.f E and an internal resistance r. The value of E is the open-circuit voltage at the terminals and r is equal to the resistance of the network measured at the terminals, looking back into the network, with the load disconnected and all sources removed and replaced by their internal resistances.

Refer to Fig. 2.0-2(a), (b), (c) and (d)

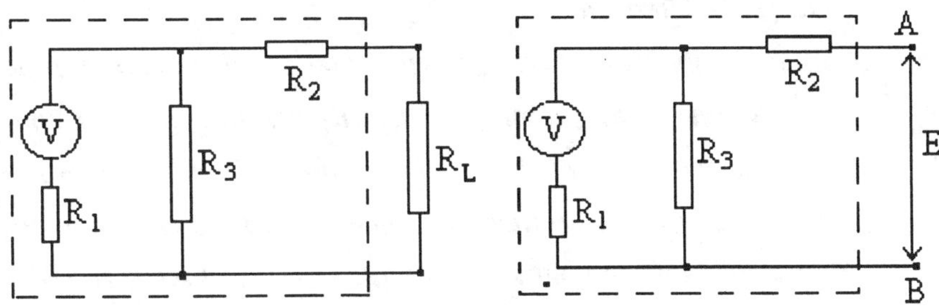

(a) Active two-terminal network   (b) Open-circuit voltage E at terminals AB

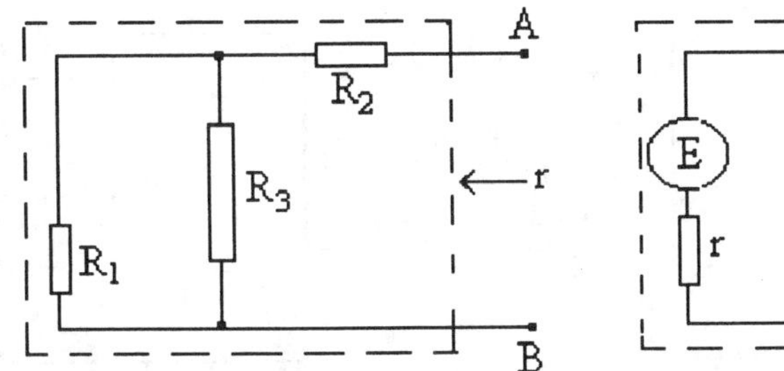

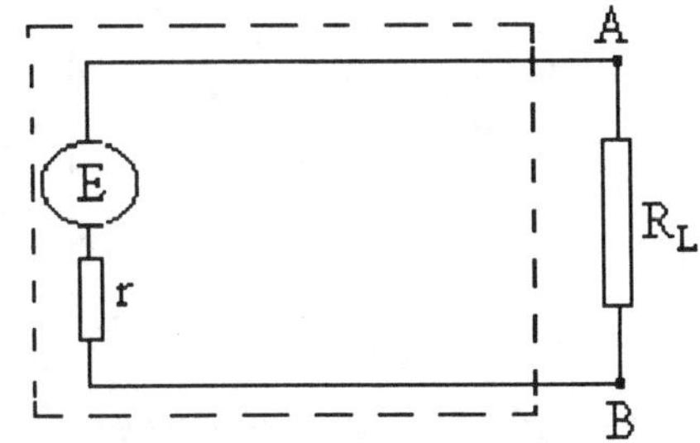

(c) All sources of e.m.f removed   (d) Constant-voltage generator with
and r measured looking backward   e.m.f E and internal resistance r

Fig. 2.0-2

(contd)

<u>Refer to Fig. 2.0-2(b)</u>

$$\text{Voltage across terminals } AB = V \times \frac{R_3}{R_1 + R_3} \qquad \text{(Eq. 2.0-0)}$$

<u>Refer to Fig. 2.0-2(c)</u>

$$\text{Resistance } r = R_2 + \frac{R_1 R_3}{R_1 + R_3} \qquad \text{(Eq. 2.0-1)}$$

<u>Refer to Fig. 2.0-2(d)</u>

$$\text{Constant-voltage generator with e.m.f } E = \frac{V R_3}{R_1 + R_3} \qquad \text{[Same as Eq. 2.0-0]}$$

$$\text{Generator } E \text{ with an internal resistance } r = R_2 + \frac{R_1 R_3}{R_1 + R_3} \qquad \text{[Same as Eq. 2.0-1]}$$

The complex circuit of Fig. 2.0-2(a) is now reduced to a constant-voltage generator E having an internal resistance r and the load R_L connected across the terminals of the generator seen in Fig. 2.0-2(d).

Norton's Theorem

An active network having a pair of terminals may be replaced by an equivalent circuit having a constant-current generator equivalent to the short-circuit current (I_{sc}) at the terminals in parallel with the internal resistance r equal to the resistance seen at the terminals looking backward into the network with all sources of e.m.f removed and replaced by their internal resistances.

The theorem may be illustrated by the diagrams in Fig. 2.0-3 (a) to (d).

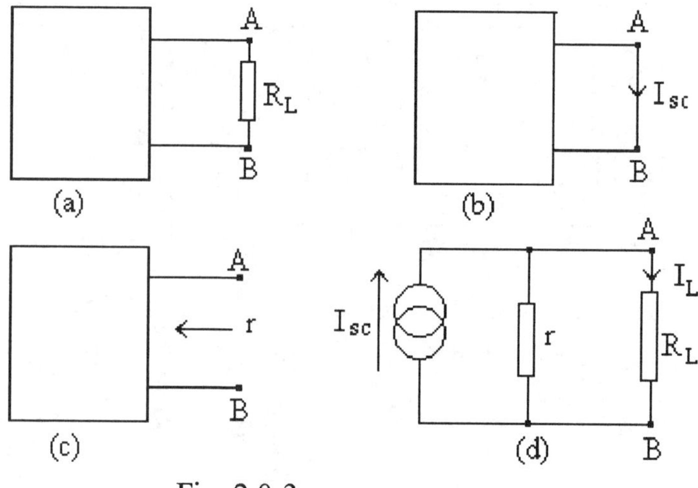

Fig. 2.0-3

(contd - Norton's Theorem)

<u>Refer to Figs. 2.3-0 (a), (b), (c) and (d)</u>

(a)   represent an active two-terminal network

(b)   display the active two-terminal network with short-circuit current, I_{sc}

(c)   illustrate the internal resistance r measured looking back into the network (indicated by arrow); all sources of e.m.f removed and replaced by their internal resistances

(d)   show the final circuit with equivalent constant-current generator I_{sc} and internal resistance r

$$\text{Current in load, } I_L = \frac{I_{sc} \times r}{R_L + r}$$

Delta - star (also known as π - T) Transformation

If two circuits are to be identical, then their resistances must be equal. If the resistances in a delta (π) connected circuit are to be equal to the resistances in a star (T) connected circuit, then the resistance between a pair of terminals in the star circuit, with the third terminal open, must be equal to the resistance measured across the identical pair of terminals in the delta circuit.

<u>Consider the delta (π) to star (T) transformation in Fig. 2.0-4(a) and (b)</u>

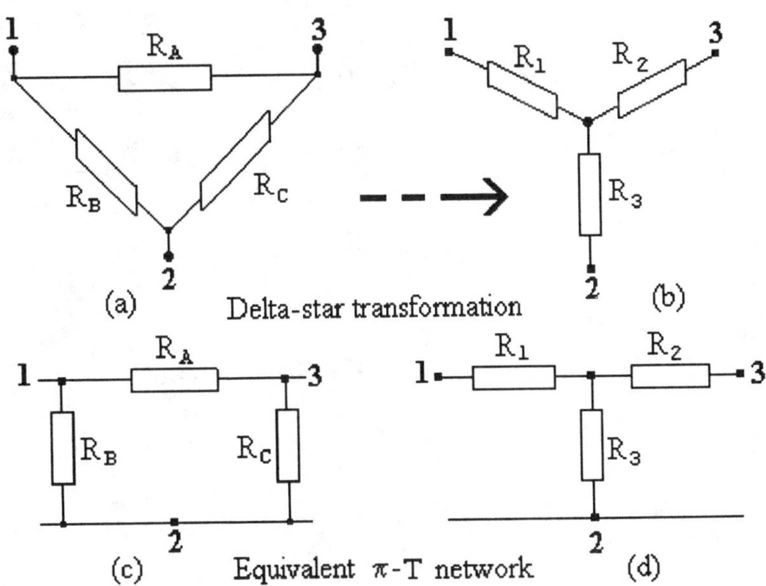

Fig. 2.0-4

(contd - delta-star transformation)

The equivalent π- T networks are shown in Figs. 2.0-4 (c) and (d).
By delta-star transformation, we mean replacing the delta network
by an equivalent star network. Consequently, the object is to express
R_1, R_2 and R_3 in terms of R_A, R_B and R_C.

The resistance between terminals 1 and 2 is

$$R_1 + R_3 = \frac{R_B (R_A + R_C)}{R_A + R_B + R_C} \qquad \text{(Eq. 2.0-2)}$$

The resistance between terminals 2 and 3 is

$$R_2 + R_3 = \frac{R_C (R_A + R_B)}{R_A + R_B + R_C} \qquad \text{(Eq. 2.0-3)}$$

The resistance between terminals 3 and 1 is

$$R_1 + R_2 = \frac{R_A (R_B + R_C)}{R_A + R_B + R_C} \qquad \text{(Eq. 2.0-4)}$$

Subtracting Eq. 2.0-3 from Eq. 2.0-2, we get

$$(R_1 + R_3) - (R_2 + R_3) = \frac{R_B (R_A + R_C) - R_C (R_A + R_B)}{R_A + R_B + R_C}$$

$$R_1 + R_3 - R_2 - R_3 = \frac{R_A R_B + R_B R_C - R_A R_C - R_B R_C}{R_A + R_B + R_C}$$

$$R_1 - R_2 = \frac{R_A (R_B - R_C)}{R_A + R_B + R_C} \qquad \text{(Eq. 2.0-5)}$$

Adding Eq. 2.0-4 and Eq. 2.0-5, we get

$$(R_1 + R_2) + (R_1 - R_2) = \frac{R_A (R_B + R_C) + R_A (R_B - R_C)}{R_A + R_B + R_C}$$

$$R_1 + R_2 + R_1 - R_2 = \frac{R_A R_B + R_A R_C + R_A R_B - R_A R_C}{R_A + R_B + R_C}$$

$$2R_1 = \frac{2 R_A R_B}{R_A + R_B + R_C}$$

$$R_1 = \frac{R_A R_B}{R_A + R_B + R_C} \qquad \text{(Eq. 2.0-6)}$$

Similarly

$$R_2 = \frac{R_A R_C}{R_A + R_B + R_C} \qquad \text{(Eq. 2.0-7)}$$

and

$$R_3 = \frac{R_B R_C}{R_A + R_B + R_C} \qquad \text{(Eq. 2.0-8)}$$

Star-delta transformation (T - π)

<u>Refer to Fig. 2.0-5(a) and (b)</u>

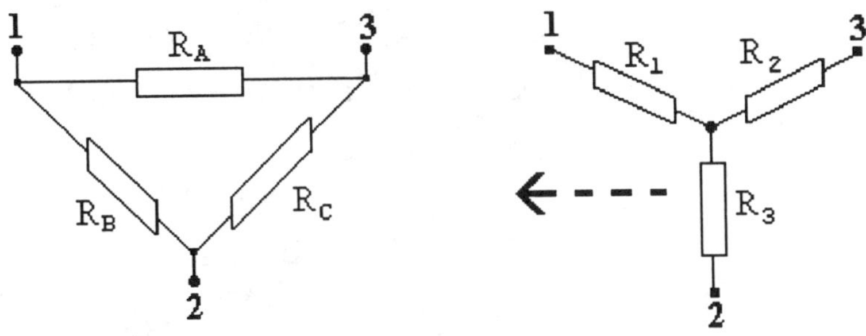

(a)   Star-delta transformation (b)

Fig. 2.0-5

To simplify calculation it is convenient to express the resistances in terms of conductances. Thus the conductance $G_A = {}^1\!/R_A$, $G_1 = {}^1\!/R_1$, etc.

The input conductance between terminals 1 and 2 with terminals 3 and 2 short-circuited is

$$G_A + G_B = \frac{G_1\,(G_2 + G_3)}{G_1 + G_2 + G_3} \qquad \text{(Eq. 2.0-9)}$$

The input conductance between terminals 2 and 3 with terminals 1 and 2 short-circuited is

$$G_A + G_C = \frac{G_2\,(G_1 + G_3)}{G_1 + G_2 + G_3} \qquad \text{(Eq. 2.0-10)}$$

The input conductance between terminals 1 and 2 with terminals 1 and 3 short-circuited is

$$G_B + G_C = \frac{G_3\,(G_1 + G_2)}{G_1 + G_2 + G_3} \qquad \text{(Eq. 2.0-11)}$$

Equate the above equations to obtain two simultaneous equations and solve for G_A or G_B or G_C. In this introduction, G_A is chosen.

Subtracting Eq. 2.0-11 from Eq. 2.0-9, we get

$$(G_A + G_B) - (G_B + G_C) = \frac{G_1\,(G_2 + G_3) - G_3\,(G_1 + G_2)}{G_1 + G_2 + G_3}$$

$$G_A + G_B - G_B - G_C = \frac{G_1 G_2 + G_1 G_3 - G_1 G_3 - G_2 G_3}{G_1 + G_2 + G_3}$$

$$G_A - G_C = \frac{G_1 G_2 - G_2 G_3}{G_1 + G_2 + G_3} \qquad \text{(Eq. 2.0-12)}$$

(contd)

Adding Eq. 2.0-10 and Eq. 2.0-12, we get

$$(G_A + G_C) + (G_A - G_C) = \frac{G_1\,G_2 + G_2\,G_3 + G_1\,G_2 - G_2\,G_3}{G_1 + G_2 + G_3}$$

$$2\,G_A = 2G_1\,G_2/(G_1 + G_2 + G_3)$$

$$G_A = G_1\,G_2/(G_1 + G_2 + G_3)$$

$$^1/R_A = \frac{1}{G_1\,G_2/(G_1 + G_2 + G_3)}$$

$$R_A = \frac{G_1 + G_2 + G_3}{G_1\,G_2}$$

$$= \frac{^1/R_1 + \,^1/R_2 + \,^1/R_3}{^1/(R_1\,R_2)}$$

$$= \frac{R_2\,R_3 + R_1\,R_3 + R_1\,R_2}{R_1\,R_2\,R_3} \div \,^1/(R_1\,R_2)$$

$$= \frac{R_2\,R_3 + R_1\,R_3 + R_1\,R_2}{R_1\,R_2\,R_3} \times R_1\,R_2$$

That is,
$$R_A = \frac{R_1\,R_2 + R_1\,R_3 + R_2\,R_3}{R_3} \qquad \text{(Eq. 2.0-13)}$$

Similarly

$$R_B = \frac{R_1\,R_2 + R_1\,R_3 + R_2\,R_3}{R_2} \qquad \text{(Eq. 2.0-14)}$$

$$R_C = \frac{R_1\,R_2 + R_1\,R_3 + R_2\,R_3}{R} \qquad \text{(Eq. 2.0-15)}$$

Summarizing the delta-star and star-delta transformations

Delta to star (π- T)

$$R_1 = \frac{R_A\,R_B}{R_A + R_B + R_C}$$

$$R_2 = \frac{R_A\,R_C}{R_A + R_B + R_C}$$

$$R_3 = \frac{R_B\,R_C}{R_A + R_B + R_C}$$

Star to delta (T- π)

$$R_A = \frac{R_1\,R_2 + R_1\,R_3 + R_2\,R_3}{R_3}$$

$$R_B = \frac{R_1\,R_2 + R_1\,R_3 + R_2\,R_3}{R_2}$$

$$R_C = \frac{R_1\,R_2 + R_1\,R_3 + R_2\,R_3}{R_1}$$

Expressing the summary in words

Delta to star transformation

To replace a delta network by a star network, the resistance or impedance of the star network is equal to the product of the adjacent two delta resistances divided by the sum of the delta resistances.

(contd - Expressing the summary in words)

<u>Refer to Fig. 2.0-6</u>

R_1 is given by the product of the two adjacent delta resistances, $R_A R_B$, divided by the sum of the delta resistances, $R_A + R_B + R_C$.

Star to delta transformation

To replace a star network by a delta network, the resistance of the delta network is equal to the sum of the products of all pairs of resistances in the star divided by the opposite resistance of the pair in the star network.

<u>Refer to Fig. 2.0-6</u>

R_A is given by the sum of the three products $R_1 R_2 + R_1 R_3 + R_2 R_3$ divided by R_3, being the opposite resistance of the star network.

Fig. 2.0-6 is for visual convenience.

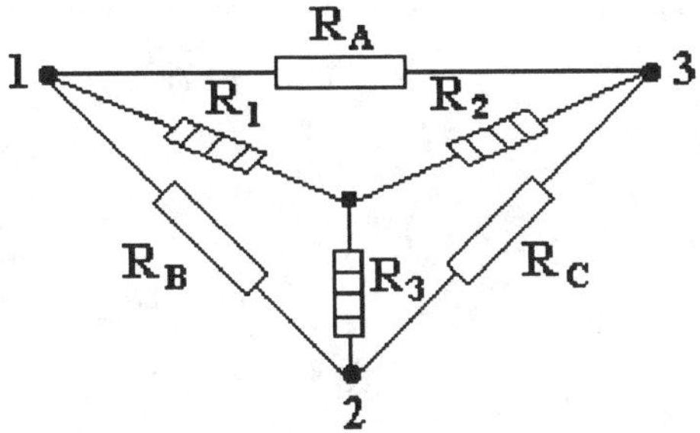

Star network enclosed in delta network for visual convenience
and simplicity in star-delta or delta-star transformation

Fig. 2.0-6

Maximum Power Transfer Theorem

This theorem states that the power transfer to a load is at maximum when the load resistance is equal to the source internal resistance.

When this condition is achieved, it is referred to as the *matched* condition.

<u>Consider the diagram in Fig. 2.0-7</u>

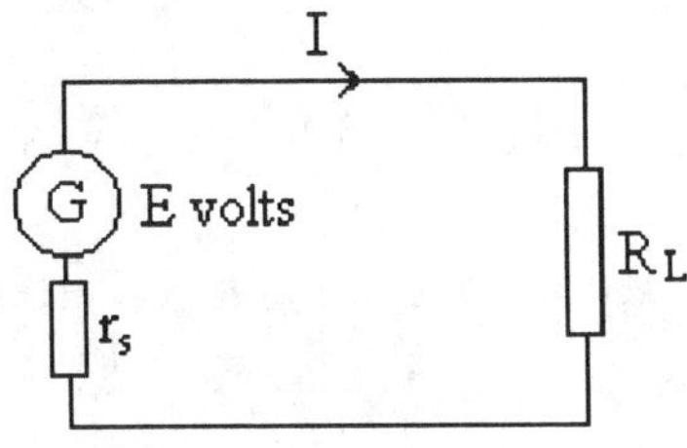

Fig. 2.0-7

(contd)

The generator G having an internal resistance r_s supplies a load R_L.

$$\text{Current } I = \frac{E}{r_s + R_L}$$

$$\text{Power P consumed by } R_L = I^2 R_L = \frac{E^2}{(r_s + R_L)^2} \times R_L$$

$$= \frac{E^2 R_L}{r_s^2 + 2R_L r_s + R_L^2}$$

$$= \frac{E^2}{(r_s^2/R_L) + 2r_s + R_L} \qquad \text{(Eq. 2.0-16)}$$

In order to achieve maximum power transfer, the denominator of Eq. 2.0-16 must be a minimum.

This condition is determined by differentiating the denominator with respect to R_L and equating to zero. Thus

$$\frac{d}{dR_L}[(r_s^2/R_L) + 2r_s + R_L] = 0$$

$$\frac{d}{dR_L}[r_s^2 R_L^{-1} + 2r_s + R_L] = 0$$

$$-1\, r_s^2 R_L^{-2} + 1 = 0$$

$$-(r_s^2/R_L^2) + 1 = 0$$

$$1 = r_s^2/R_L^2$$

$$R_L^2 = r_s^2$$

$$\therefore \qquad \mathbf{R_L = r_s} \qquad \text{(Eq. 2.0-17)}$$

Hence, when R_L (load resistance) $= r_s$ (source internal resistance), maximum power transfer is achieved.

Example 2.1

A d.c generator supplies 20 A to four resistive loads having resistances of 4 Ω, 8 Ω, 10 Ω and 20 Ω connected in parallel.

Calculate the current taken by each load.

Solution

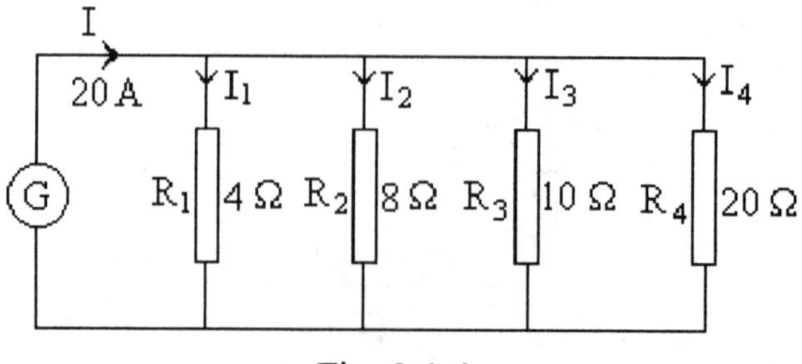

Fig. 2.1-0

(contd)

<u>Refer to Fig. 2.1-0</u>

$$\text{E.m.f of generator} = I\,R_T$$

$$^1\!/R_T = {}^1\!/R_1 + {}^1\!/R_2 + {}^1\!/R_3 + {}^1\!/R_4$$

$$= {}^1\!/4 + {}^1\!/8 + {}^1\!/10 + {}^1\!/20$$

$$= \frac{10 + 5 + 4 + 2}{40} \qquad = \frac{21}{40}$$

$$R_T = {}^{40}\!/21$$

$$\therefore \quad \text{E.m.f} = 20 \times {}^{40}\!/21 \qquad = \underline{38.1\ V}$$

and $\quad$ Current $I_1 = {}^V\!/R_1 = {}^{38.1}\!/4 = \mathbf{\underline{9.52\ A}}$

$\qquad\quad$ Current $I_2 = {}^V\!/R_2 = {}^{38.1}\!/8 = \mathbf{\underline{4.76\ A}}$

$\qquad\quad$ Current $I_3 = {}^V\!/R_3 = {}^{38.1}\!/10 = \mathbf{\underline{3.81\ A}}$

$\qquad\quad$ Current $I_4 = {}^V\!/R_4 = {}^{38.1}\!/20 = \mathbf{\underline{1.91\ A}}$

Alternatively

The parallel resistive loads can be expressed as conductances.
Thus $\quad {}^1\!/R_1 = G_1, \ {}^1\!/R_2 = G_2, \ {}^1\!/R_3 = G_3$, and ${}^1\!/R_4 = G_4$.

By applying Kirchhoff's First Law (current law), we get

$$I_1 + I_2 + I_3 + I_4 = I_T$$

$$I_T = V(G_1 + G_2 + G_3 + G_4)$$

$$= V\,G_T$$

$$V = {}^{I_T}\!/G_T$$

and $\quad I_1 = V\,G_1$

$$= I_T\,{}^{G_1}\!/G_T$$

Similarly $\quad I_2 = I_T\,{}^{G_2}\!/G_T$

$$I_3 = I_T\,{}^{G_3}\!/G_T$$

$$I_4 = I_T\,{}^{G_4}\!/G_T$$

Now $\quad G_1 = {}^1\!/4 = \underline{0.25\ S}$

$\qquad\quad G_2 = {}^1\!/8 = \underline{0.125\ S}$

$\qquad\quad G_3 = {}^1\!/10 = \underline{0.1\ S}$

$\qquad\quad G_4 = {}^1\!/20 = \underline{0.05\ S}$

$$\Sigma G = G_T = 0.25 + 0.125 + 0.1 + 0.05$$

$$= \underline{0.525\ S}$$

(contd)

$$\therefore \quad I_1 = \frac{20 \times 0.25}{0.525} \qquad = \underline{\textbf{0.952 A}}$$

$$I_2 = \frac{20 \times 0.125}{0.525} \qquad = \underline{\textbf{4.76 A}}$$

$$I_3 = \frac{20 \times 0.1}{0.525} \qquad = \underline{\textbf{3.81 A}}$$

$$I_4 = \frac{20 \times 0.05}{0.525} \qquad = \underline{\textbf{1.91 A}}$$

Example 2.2

Three resistive loads of 4 Ω, 6 Ω, and 12 Ω are connected in parallel and the combination is connected to a 10 Ω load in series with a generator having an e.m.f of 120 V and an internal resistance of 0.2 Ω. Calculate the terminal voltage of the generator and the current in each load.

Solution

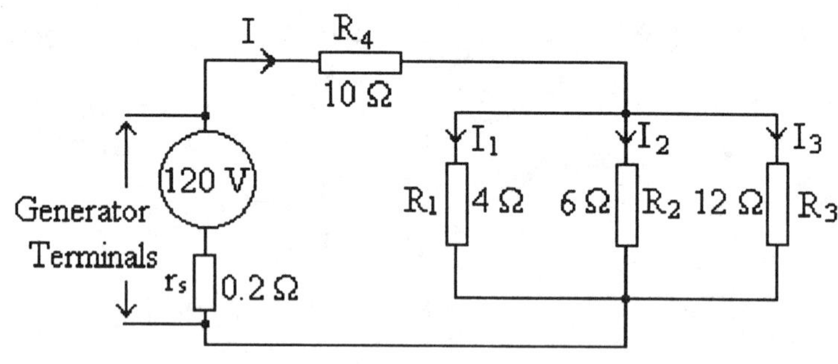

Fig. 2.2-0

<u>Refer to Fig. 2.2-0</u>

Equivalent resistance of parallel loads is

$$\frac{1}{R_p} = \frac{1}{R_1} + \frac{1}{R_2} + \frac{1}{R_3}$$

$$= \frac{1}{4} + \frac{1}{6} + \frac{1}{12}$$

$$= \frac{3 + 2 + 1}{12} = \underline{\frac{6}{12}}$$

$$\therefore \quad R_p = \frac{12}{6} = \underline{2\,\Omega}$$

Total circuit resistance, $R_T = R_4 + R_p + r_s$

$$= 10 + 2 + 0.2 \qquad = \underline{12.2\,\Omega}$$

Current flowing in circuit, $I = \frac{\text{e.m.f}}{R_T} = \frac{120}{12.2} = \underline{9.836\ A}$

(contd)

$$\text{That is, current through } R_4 = I = \mathbf{\underline{9.836\ A}}$$

$$\text{Internal voltage drop in generator} = I\,r_s$$
$$= 9.836 \times 0.2 \qquad = \underline{1.9672\ V}$$

$$\text{Generator's terminal voltage} = \text{e.m.f} - \text{internal voltage drop}$$
$$= 120 - 1.9672 \qquad = \mathbf{\underline{118.0328\ V}}$$

$$\text{P.d across parallel branch, } V_p = I \times R_p$$
$$= 9.836 \times 2 \qquad = \underline{19.672\ V}$$

$\therefore$ Current distribution in each of the parallel loads is

$$I_1 = V_p/R_1 \qquad = \frac{19.672}{4} \qquad = \mathbf{\underline{4.918\ A}}$$

$$I_2 = V_p/R_2 \qquad = \frac{19.672}{6} \qquad = \mathbf{\underline{3.279\ A}}$$

$$I_3 = V_p/R_3 \qquad = \frac{19.672}{12} \qquad = \mathbf{\underline{1.639\ A}}$$

Example 2.3

Two noninductive resistors of 250Ω are connected in series across a constant-voltage source. Determine the value of a resistor which when connected in parallel with one of the resistances will reduce the p.d across its terminals by 1%.

Solution

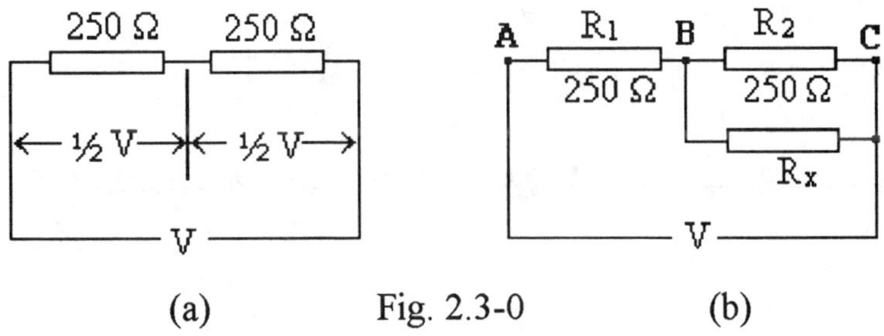

Fig. 2.3-0

Refer to Fig. 2.3-0(a)

$$\text{Let voltage } V = \text{unity (1)}$$

Then voltage across either of the 250 Ω resistors is
$$= V/2 \text{ volts}$$

Refer to Fig. 2.3-0(b)

The voltage across parallel combination $R_2\,R_x$ is

$$V_{BC} = V/2 - (1\% \text{ of } V/2)$$

(contd)

$$= 0.5\text{ V} - (0.01 \times 0.5\text{ V})$$
$$= 0.5\text{ V} - 0.005\text{ V} \qquad = \underline{0.495\text{ V}}$$

Hence, voltage across R_1 is

$$V_{AB} = (1 - 0.495)\text{ V} \qquad = \underline{0.505\text{ V}}$$

Ratio of $R_{AB} : R_{BC} \qquad = $ ratio of $0.505 : 0.495$

That is, $R_{AB}/R_{BC} \qquad = \dfrac{0.505}{0.495}$

$$\dfrac{250}{R_{BC}} = \dfrac{0.505}{0.495}$$

$$R_{BC} = \dfrac{250 \times 0.495}{0.505} \qquad = \underline{245.05\ \Omega}$$

Refer to Fig. 2.3-0(b)

Resistance of parallel combination is

$$R_{BC} = \dfrac{R_2 R_x}{R_2 + R_x}$$

$$245.05 = \dfrac{250\, R_x}{250 + R_x}$$

$$245.05\,(250 + R_x) = 250\, R_x$$
$$(245.05 \times 250) + 245.05\, R_x = 250\, R_x$$
$$250\, R_x - 245.05\, R_x = 245.05 \times 250$$
$$4.95\, R_x = 61262.5$$
$$R_x = \dfrac{61262.5}{4.95} \qquad = \underline{12376.26\ \Omega}$$

Hence resistor to be connected in parallel is approximately 12.38 kΩ

Example 2.4

Two groups of resistances are connected in series across a 120 V supply.
Group 1 consists of three resistors, 40 Ω, 60 Ω, and 80 Ω, in parallel.
Group 2 consists of two resistors, 30 Ω and 60 Ω, in parallel.
Calculate:

    (a)   the resistance of the entire circuit

    (b)   the p.d across each group

    (c)   the current in each resistance

(contd)

Solution

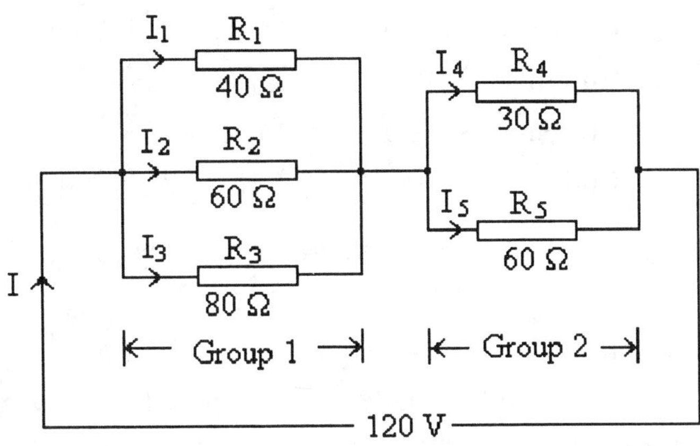

Fig. 2.4-0

<u>Refer to Fig. 2.4-0</u>

Equivalent resistance of

Group 1 $1/R_{G_1} = 1/R_1 + 1/R_2 + 1/R_3$

$$= 1/40 + 1/60 + 1/80$$

$$= \frac{6 + 4 + 3}{240} = \frac{13}{240}$$

$\therefore \quad R_{G_1} = 240/13 = \underline{18.46\ \Omega}$

Group 2 $R_{G_2} = \frac{R_4 R_5}{R_4 + R_5} = \frac{30 \times 60}{30 + 60}$

$$= \underline{20\ \Omega}$$

(a)   Total circuit resistance is

$$R_T = R_{G_1} + R_{G_2}$$

$$= 18.46 + 20 \qquad = \underline{\mathbf{38.46\ \Omega}}$$

(b)   Current flowing in the circuit is

$$I = V/R_T$$

$$= 120/38.46 \qquad = \underline{3.12\ A}$$

P.d across Group 1 $= I\,R_{G_1}$

That is, $V_1 = 3.12 \times 18.46 \qquad = \underline{\mathbf{57.60\ V}}$

P.d across Group 2 $= V - 57.60$

That is, $V_2 = 120 - 57.60 \qquad = \underline{\mathbf{62.40\ V}}$

(contd)

(c)

$$\text{Current in } R_1 = I_1 = V_1/R_1 = 57.60/40 = \mathbf{1.44\ A}$$
$$\text{Current in } R_2 = I_2 = V_1/R_2 = 57.60/60 = \mathbf{\underline{0.96\ A}}$$
$$\text{Current in } R_3 = I_3 = V_1/R_3 = 57.60/80 = \mathbf{\underline{0.72\ A}}$$
$$\text{Current in } R_4 = I_4 = V_2/R_4 = 62.40/30 = \mathbf{\underline{2.08\ A}}$$
$$\text{Current in } R_5 = I_5 = V_2/R_5 = 62.40/60 = \mathbf{\underline{1.04\ A}}$$

Verifying answers by using Kirchhoff's Law (current law), we get

$$\Sigma I = 3.12\ A$$

Thus for Group 1	$I_1 + I_2 + I_3$	$= 3.12\ A$
	$1.44 + 0.96 + 0.72$	$= 3.12\ A$ (verified)
and for Group 2	$I_3 + I_4$	$= 3.12\ A$
	$2.08 + 1.04$	$= 3.12\ A$ (verified)

Example 2.5

A moving-coil instrument has a resistance of 5 Ω and requires a current of 100 mA to give full-scale deflection (f.s.d). The instrument is to be adapted as an ammeter and then as a voltmeter. Calculate the resistances necessary to calibrate the instrument to read as

> *(a)   an ammeter, 0 – 50 A (f.s.d)*
> *(b)   a voltmeter, 0 – 500 V (f.s.d)*

Solution

Whether the instrument is used as an ammeter or a voltmeter, the maximum permissible current through the meter is 100 mA. To calibrate as an ammeter, a *shunt* resistor (R_{shunt}) is connected in parallel across the terminals of the instrument such that only 0.1 A is permitted to pass through the meter while the excess current is shunted away.

<u>Refer to Fig. 2.5-0(a)</u>

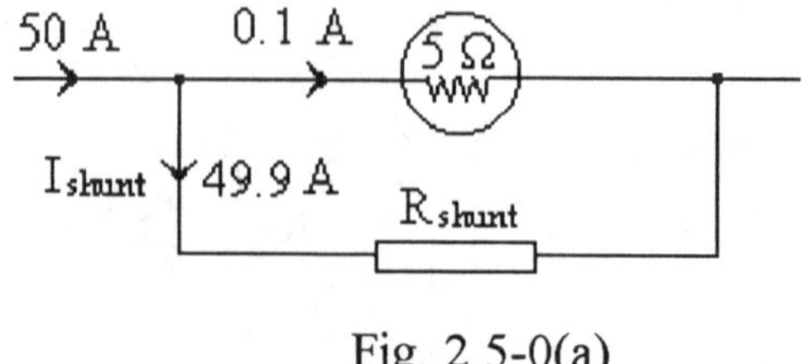

Fig. 2.5-0(a)

(contd)

(a)   *The instrument as an ammeter*

$$
\begin{aligned}
\text{Current through instrument} \quad &= 100 \text{ mA} \quad &= \underline{0.1 \text{ A}} \\
\text{Current to calibrate} \quad &= 50 \text{ A} \\
\text{Current to be shunted, } I_{sh.} \quad &= 50 - 0.1 \quad &= \underline{49.9 \text{ A}} \\
\text{Voltage across instrument} \quad &= 0.1 \times 5 \quad &= \underline{0.5 \text{ V}} \\
\text{Voltage across shunt resistor} \quad &= \text{voltage across instrument} \\
\therefore \quad \text{Value of shunt resistor, } R_{shunt} \quad &= {}^{0.5}/49.9 \quad &= \underline{\mathbf{0.01002\ \Omega}}
\end{aligned}
$$

The instrument as a voltmeter

To calibrate the instrument to be used as a voltmeter, a resistance is connected in *series* (R_{series}) with the instrument such that only the permissible 0.1 A passes through it, thus sustaining the allowable voltage drop across the instrument, while the excess voltage is dropped across the series resistance.

Refer to Fig. 2.5-0(b)

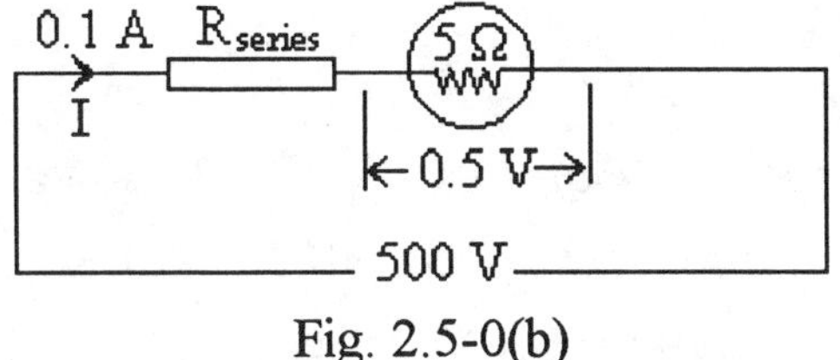

Fig. 2.5-0(b)

$$
\begin{aligned}
\text{Current through } R_{series} \quad &= 0.1 \text{ A} \\
\text{Voltage to drop by } R_{series} \quad &= 500 - 0.5 \quad &= \underline{499.5 \text{ V}} \\
\text{Value of resistance, } R_{series} \quad &= \frac{V_{Rseries}}{I} \\
&= \frac{499.5}{0.1} \quad &= \underline{\mathbf{4995\ \Omega}}
\end{aligned}
$$

Alternatively

$$
\begin{aligned}
\text{Total circuit resistance} \quad &= {}^{V}/I \\
&= {}^{500}/0.1 \quad &= \underline{5000\ \Omega} \\
\text{Resistance of instrument} \quad &= 5 \ \Omega \\
\therefore \quad \text{Series resistance} \quad &= 5000 - 5 \quad &= \underline{\mathbf{4995\ \Omega}}
\end{aligned}
$$

Example 2.6

Three cells, each having an e.m.f of 2 V and negligible resistance, are connected in series. A resistor of 100 Ω and one of 25 Ω are connected in series across the battery, the 100 Ω resistor being connected to the positive terminal. Calculate the current flowing in a 10 Ω resistor connected between a tapping on the battery 4 V from the positive end and the junction of the two resistors.

Solution

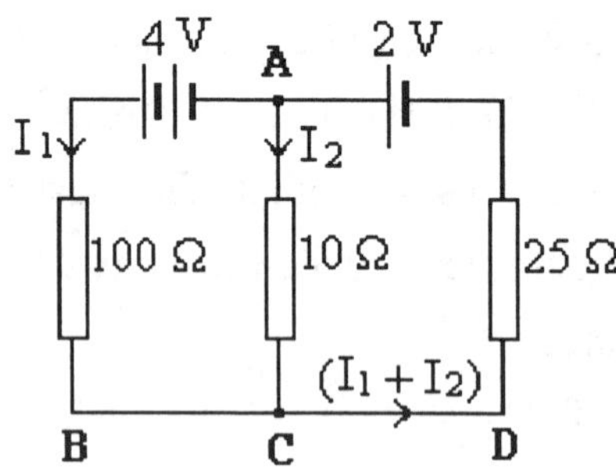

Fig. 2.6-0

Applying Kirchhoff's Law to Fig. 2.6-0, we get

Closed loop ABCA: $\quad 100\,I_1 - 10\,I_2 \qquad\qquad = 4 \qquad$ (Eq. 2.6-0)

Closed loop ACDA: $\quad 10\,I_2 + 25(I_1 + I_2) \quad = 2$

$$25\,I_1 + 35\,I_2 \qquad = 2 \qquad \text{(Eq. 2.6-1)}$$

Subtract Eq. 2.6-1 from Eq. 2.6-0 to solve for I_2

(Eq. 2.6-0) x 1 $\qquad 100\,I_1 - 10\,I_2 \qquad\quad = 4$

(Eq. 2.6-1) x 4 $\qquad \underline{100\,I_1 + 140\,I_2 \qquad\quad = 8}$

$$-\,150\,I_2 \qquad\quad = -4$$

Current through 10 Ω = I_2 = $4/150$ = **0.02667 A** = **26.67 mA**

Example 2.7

A Wheatstone bridge is supplied by a 2 V battery of negligible resistance. The resistances of the arms of the bridge are shown in Fig. 2.7-0. Determine the value and direction of the current in the galvanometer circuit using.

(a)   Kirchhoff's Laws

(b)   Thevenin's Theorem

(contd)

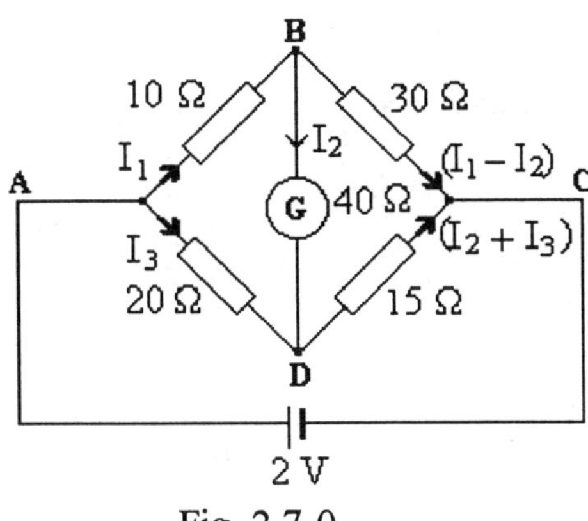

Fig. 2.7-0

Solution

(a) **By Kirchhoff's Laws** (<u>Refer to Fig. 2.7-0</u>)

Closed loop **ABDA:** $\quad 10\,I_1 + 40\,I_2 - 20\,I_3 \qquad = 0$

Divide by 10: $\qquad\qquad I_1 + 4\,I_2 - 2\,I_3 \qquad = 0 \qquad$ (Eq. 2.7-0)

Closed loop **BCDB:** $30(I_1 - I_2) - 15(I_2 + I_3) - 40\,I_2 = 0$

$$30\,I_1 - 30\,I_2 - 15\,I_2 - 15\,I_3 - 40\,I_2 = 0$$

$$30\,I_1 - 85\,I_2 - 15\,I_3 = 0$$

Divide by 5: $\qquad\qquad 6\,I_1 - 17\,I_2 - 3\,I_3 = 0 \qquad$ (Eq. 2.7-1)

Closed loop **ADCA:** $\qquad 20\,I_3 + 15\,(I_2 + I_3) \quad = 2$

$$20\,I_3 + 15\,I_2 + 15\,I_3 = 2$$

$$15\,I_2 + 35\,I_3 = 2 \qquad \text{(Eq. 2.7-2)}$$

From Eq. 2.7-0 $\qquad -4\,I_2 + 2\,I_3 = I_1$

Substituting this value for I_1 in Eq. 2.7-1, we get

$$6\,(-4\,I_2 + 2\,I_3) - 17\,I_2 - 3\,I_3 \quad = 0$$

$$-24\,I_2 + 12\,I_3 - 17\,I_2 - 3\,I_3 \quad = 0$$

$$-41\,I_2 + 9\,I_3 \quad = 0 \qquad \text{(Eq. 2.7-3)}$$

Subtract Eq. 2.7-3 from Eq. 2.7-2

(Eq. 2.7-2) x 9 $\qquad 135\,I_2 + 315\,I_3 \quad = 18$

(Eq. 2.7-3) x 35 $\qquad \underline{-1435\,I_2 + 315\,I_3 \quad = 0}$

$$\underline{1570\,I_2 \qquad\qquad = 18}$$

$$\therefore \quad I_2 \;=\; {}^{18}/1570 \quad = \underline{\textbf{0.01146 A}}$$

$$= \underline{\textbf{11.46 mA}}$$

Note: Since the value of I_2 is positive, the direction of the current in the galvanometer circuit is that assumed in Fig. 2.7-0, that is, from **B** to **D**.

(contd)

(b)   ***By Thevenin's Theorem***

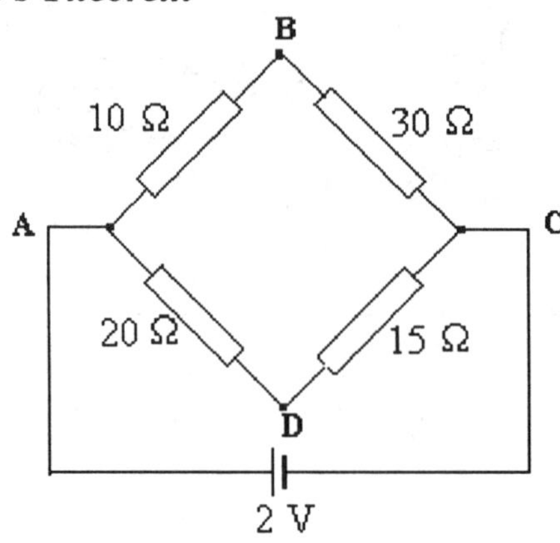

Fig. 2.7-1(a)

<u>Refer to Fig. 2.7-1(a)</u>

Step 1: Remove the galvanometer from the circuit and determine the
voltage at terminals **B** and **D**.

Voltage drop across **A** and **B** $= 2 \times \dfrac{10}{10+30} = $ <u>0.5 V</u>

$\therefore \qquad V_B = 2 - 0.5 \qquad = $ <u>1.5 V</u>

Voltage drop across **A** and **D** $= 2 \times \dfrac{20}{20+15} = $ <u>1.143 V</u>

$\therefore \qquad V_D = 2 - 1.143 \qquad = $ <u>0.857 V</u>

Voltage between **B** and **D** $= 1.5 - 0.857 = $ <u>0.643 V</u>

V_B is more positive than V_D. Hence, current flows from **B** to **D**.

Step 2: Remove source of supply and determine **r** between **B** and **D**.

<u>Refer to Fig. 2.7-1(b)</u>

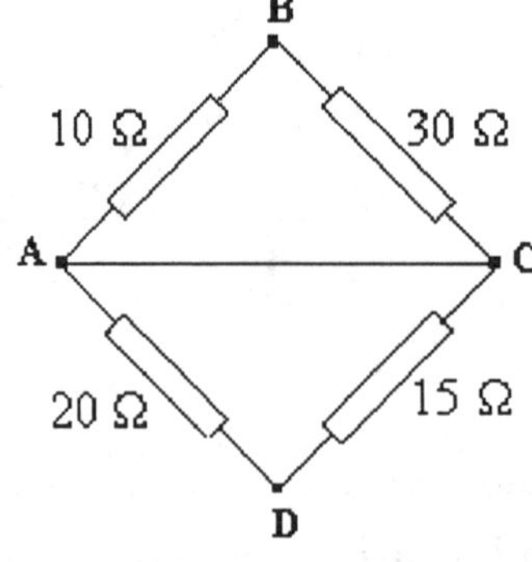

Fig. 2.7-1(b)

(contd)

The source is removed and replaced by its internal resistance. However, since its internal resistance is negligible, it is replaced by a short circuit.

$$r = \text{Equivalent resistance between BA and BC} + \text{Equivalent resistance between DA and DC}$$

$$= \frac{10 \times 30}{10 + 30} + \frac{20 \times 15}{20 + 15}$$

$$= 7.5 + 8.57 = \underline{16.07\ \Omega}$$

Step 3: Calculate the current in the galvanometer circuit from the equivalent circuit in Fig. 2.7-1(c)

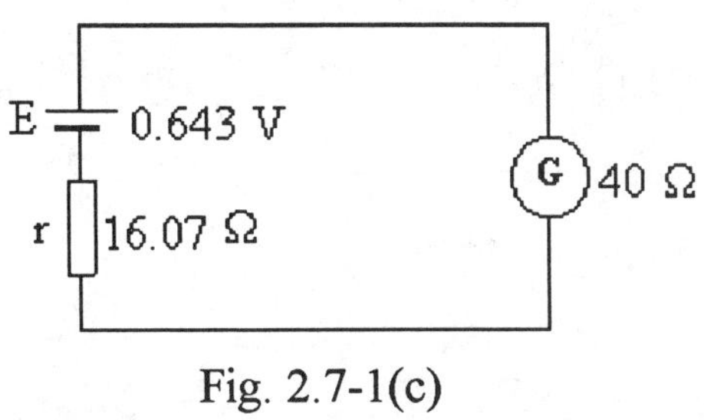

Fig. 2.7-1(c)

Current in galvanometer circuit $= E/(40 + r)$

$$= 0.643/(40 + 16.07)$$

$$= \underline{\textbf{0.01146 A}} = \underline{\textbf{11.46 mA}}$$

Example 2.8

The network shown in Fig. 2.8-0 has two sources of supply, one of 6 V and one of 4 V, with internal resistances of 2 Ω and 3 Ω, respectively. The two sources are connected in parallel across a load resistor of 6 Ω. Calculate the current in the 6 Ω resistor using the superposition theorem.
<u>*Refer to Fig. 2.8-0*</u>

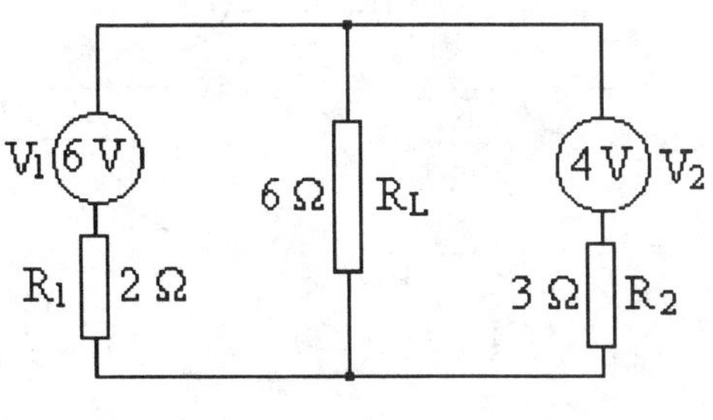

Fig. 2.8-0

59

(contd)

Solution

By superposition method

This method is best carried out in procedural steps.

Step 1: Determine the current in the load by either source acting alone, the other source being removed and replaced by its internal resistance.

Network with 6 V source acting alone [Refer to Fig. 2.8-0(a)]

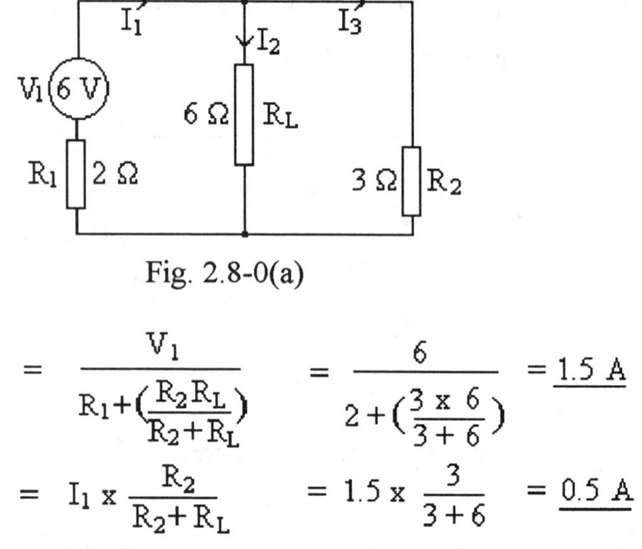

Fig. 2.8-0(a)

$$I_1 = \frac{V_1}{R_1 + \left(\dfrac{R_2 R_L}{R_2 + R_L}\right)} = \frac{6}{2 + \left(\dfrac{3 \times 6}{3 + 6}\right)} = \underline{1.5 \text{ A}}$$

$$I_2 = I_1 \times \frac{R_2}{R_2 + R_L} = 1.5 \times \frac{3}{3 + 6} = \underline{0.5 \text{ A}}$$

$$I_3 = I_1 - I_2 = 1.5 - 0.5 = \underline{1.0 \text{ A}}$$

Network with 4 V source acting alone [Refer to Fig. 2.8-0(b)]

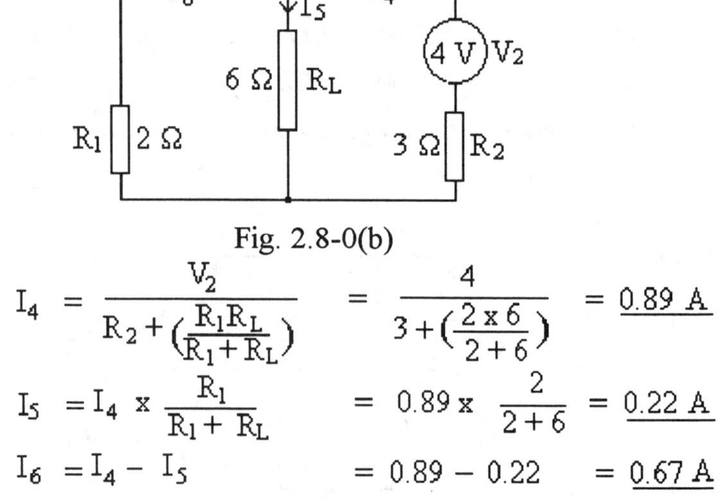

Fig. 2.8-0(b)

$$I_4 = \frac{V_2}{R_2 + \left(\dfrac{R_1 R_L}{R_1 + R_L}\right)} = \frac{4}{3 + \left(\dfrac{2 \times 6}{2 + 6}\right)} = \underline{0.89 \text{ A}}$$

$$I_5 = I_4 \times \frac{R_1}{R_1 + R_L} = 0.89 \times \frac{2}{2 + 6} = \underline{0.22 \text{ A}}$$

$$I_6 = I_4 - I_5 = 0.89 - 0.22 = \underline{0.67 \text{ A}}$$

(contd)

> ***Step 2:*** Superimpose the currents in the networks of Figs.2.8-0(a) and (b),
> redrawn as Figs. 2.8-1(a) and (b) below.
>
> Resultant current through V_1 = $I_1 - I_6$ = $1.5 - 0.67$ = <u>0.83 A</u>
>
> Resultant current through V_2 = $I_4 - I_3$ = $0.89 - 1.0$ = <u>-0.11 A</u>
>
> *The negative sign indicates that V_1 is charging V_2 at 0.11 A.*
>
> Resultant current in R_L = $I_2 + I_5$ = $0.5 + 0.22$ = <u>**0.72 A**</u>

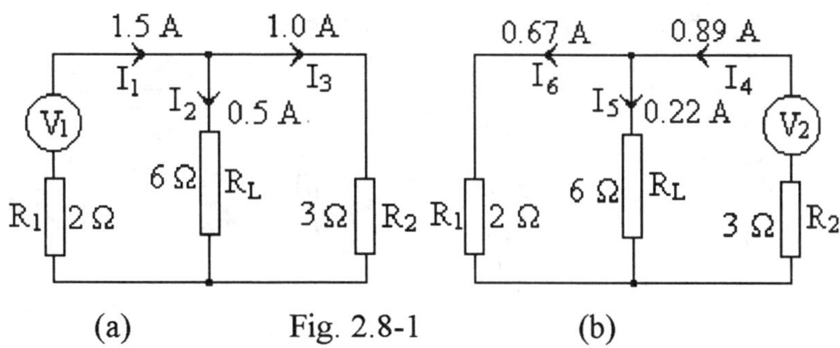

(a)   Fig. 2.8-1   (b)

Example 2.9

Calculate the current in each branch of the network shown in Fig. 2.9-0
using the superposition theorem.

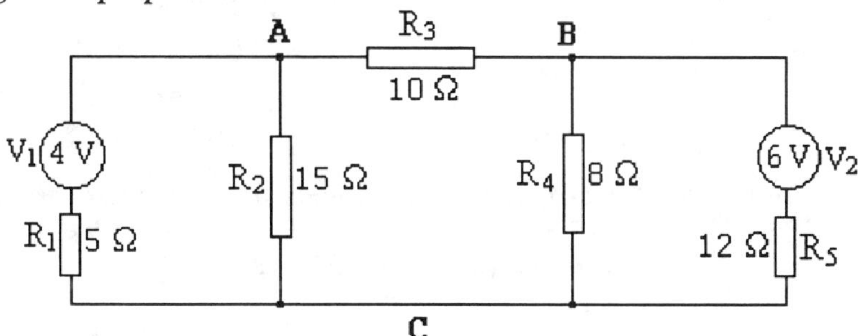

Fig. 2.9-0

Solution

4 V supply acting alone   <u>[Refer to Fig. 2.9-0(a)]</u>

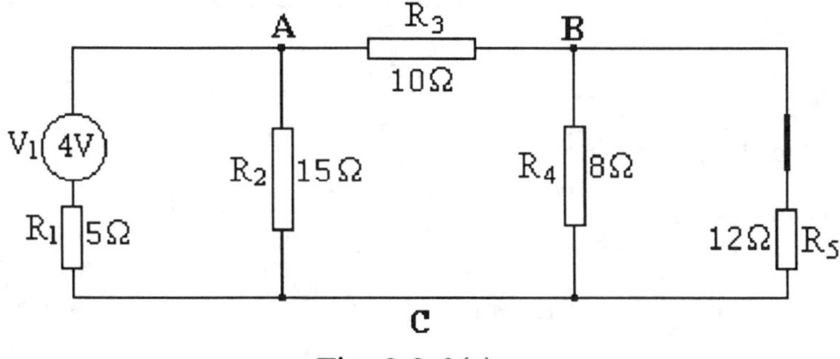

Fig. 2.9-0(a)

61

(contd)

Let resistance across **BC** $= \text{Req}_1$

$$= \frac{R_4 \times R_5}{R_4 + R_5}$$

$\therefore \qquad \text{Req}_1 = \frac{8 \times 12}{8 + 12} \qquad = \underline{4.8\ \Omega}$

The equivalent resistance of R_4 and R_5 is now replaced with a single resistance (Req 1) of 4.8 Ω.

The circuit of Fig. 2.9-0(a) is now reduced to the circuit in Fig. 2.9-0(b).

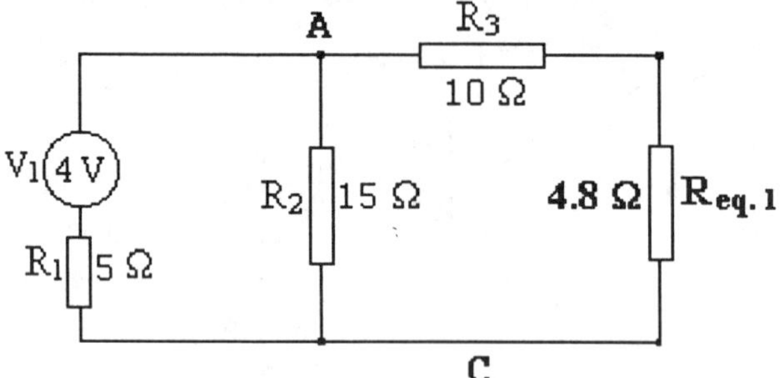

Fig. 2.9-0(b)

Refer to Fig. 2.9-0(b)

Let resistance across **AC** $= \text{Req}_2$

$$= \frac{R_2 \times (R_3 + \text{Req}_1)}{R_2 + (R_3 + \text{Req}_1)}$$

$\therefore \qquad \text{Req}_2 = \frac{15 \times (10 + 4.8)}{15 + (10 + 4.8)} \qquad = \underline{7.45\ \Omega}$

Similarly, the circuit is now reduced to the network shown in Fig. 2.9-0(c).

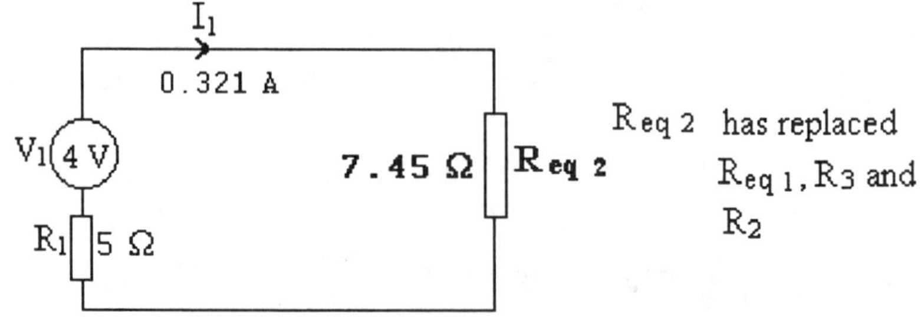

Fig. 2.9-0(c)

Refer to Fig. 2.9-0(c)

$R_T = R_1 + \text{Req}_2$

$\qquad = 5 + 7.45 \qquad = \underline{12.45\ \Omega}$

Current, $\quad I_1 = V_1 / R_T$

$\qquad = {}^4/12.45 \qquad = \underline{\mathbf{0.321\ A}}$

(contd)

Fig. 2.9-0(d) shows the current distribution with $R_{eq\,1}$ in the network.

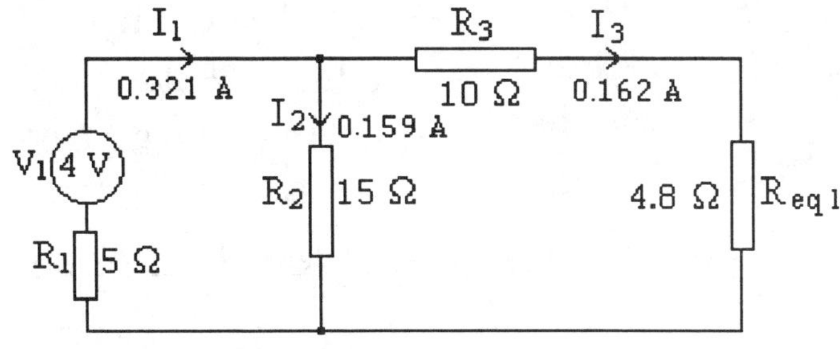

Fig. 2.9-0(d)

<u>Refer to Fig. 2.9-0(d)</u>

$$I_2 = I_1 \times \frac{R_3 + R_{eq\,1}}{R_2 + (R_3 + R_{eq\,1})}$$

$$= 0.321 \times \frac{10 + 4.8}{15 + (10 + 4.8)} = \mathbf{\underline{0.159\ A}}$$

$$I_3 = I_1 - I_2 = 0.321 - 0.159 = \mathbf{\underline{0.162\ A}}$$

<u>*Alternatively*</u>

I_3 could have been calculated first and then I_1 determined.

Thus
$$I_3 = I_1 \times {}^{R_2}/(R_2 + R_3 + R_{eq\,1})$$
$$= 0.321 \times {}^{15}/(15 + 10 + 4.8)$$
$$= 0.321 \times {}^{15}/29.8 = \mathbf{\underline{0.162\ A}}$$

$$I_2 = I_1 - I_3$$
$$= 0.321 - 0.162 = \mathbf{\underline{0.159\ A}}$$

*With **4 V supply acting alone**, the current distribution in each branch of the network is shown in Fig. 2.9-0(e).*

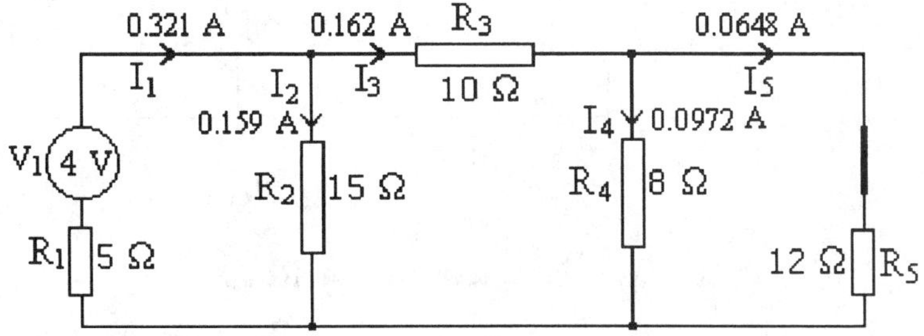

Fig. 2.9-0(e). Branch current with 4 V supply acting alone

(contd)

<u>Refer to Fig. 2.9-0(e)</u>

$$I_4 \ = \ I_3 \ \times \ R_5/(R_4 + R_5)$$
$$= \ 0.162 \ \times \ ^{12}/(8 + 12) \qquad = \ \underline{\textbf{0.0972 A}}$$
$$I_5 \ = \ I_3 - I_4 \ = \ 0.162 - 0.0972 \ = \ \underline{\textbf{0.0648 A}}$$

6 V supply acting alone [<u>Refer to Fig. 2.9-1(a)</u>]

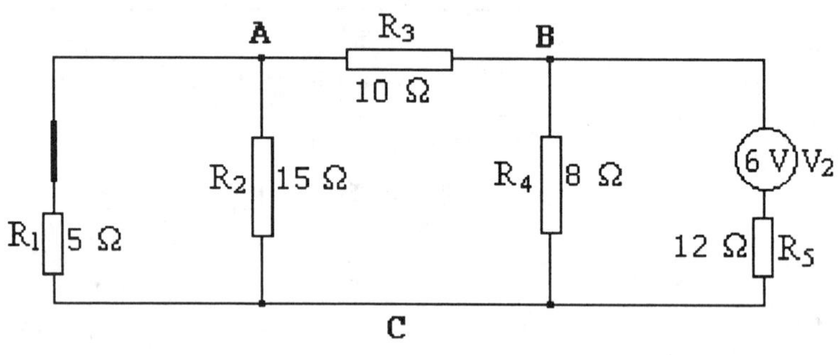

Fig. 2.9-1(a)

Let resistance across **AC** $= \ $ Req 3

$$= \ \frac{R_1 \times R_2}{R_1 + R_2}$$

$\therefore \qquad$ Req 3 $\ = \ \dfrac{5 \times 15}{5 + 15} \qquad = \ \underline{3.75 \ \Omega}$

The circuit is now reduced to the network shown in Fig. 2.9-1(b).

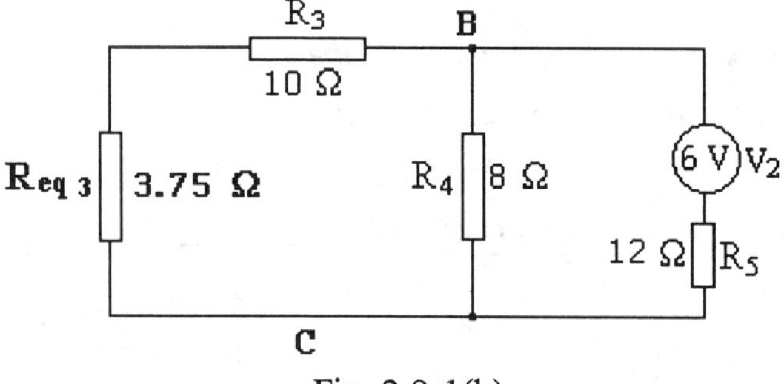

Fig. 2.9-1(b)

Let resistance across **BC** $= $ Req 4

$$= \ \frac{R_4 \times (R_3 + Req\ 3)}{R_4 + (R_3 + Req\ 3)}$$

$$= \ \frac{8 \ \times \ (10 + 3.75)}{8 \ + \ (10 + 3.75)} \ = \ \underline{5.06 \ \Omega}$$

(contd)

The circuit is now reduced to the network shown in Fig. 2.9-1(c).

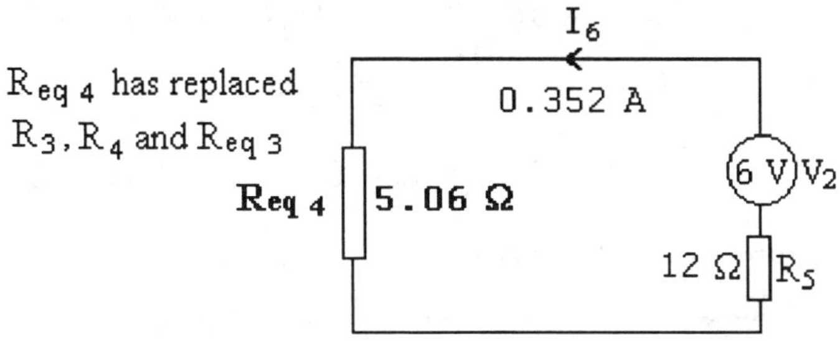

Fig. 2.9-1(c)

$$\text{Current } I_6 = V_2/(R_5 + R_{eq\,4})$$
$$= 6/(12 + 5.06) = \mathbf{0.352\ A}$$

The current distribution with Req 3 in the network is shown in Fig. 2.9-1(d).

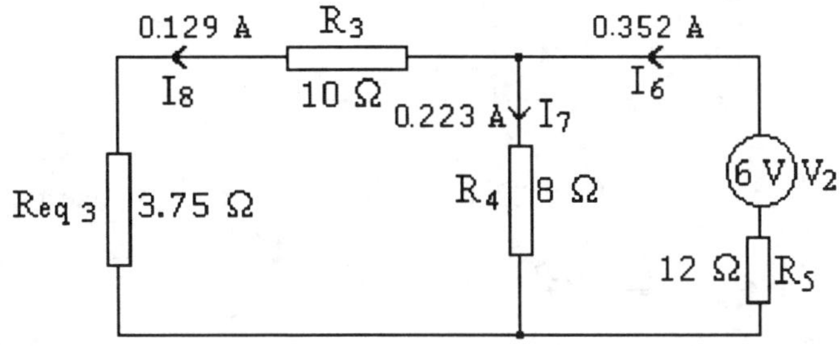

Fig. 2.9-1(d)

$$I_7 = I_6 \times \frac{R_3 + R_{eq\,3}}{(R_3 + R_{eq\,3}) + R_4}$$
$$= 0.352 \times \frac{10 + 3.75}{(10 + 3.75) + 8} = \mathbf{0.223\ A}$$
$$I_8 = I_6 - I_7 = 0.352 - 0.223 = \mathbf{0.129\ A}$$

With 6 V supply acting alone, the current in each branch of the network is shown in Fig. 2.9-1(e).

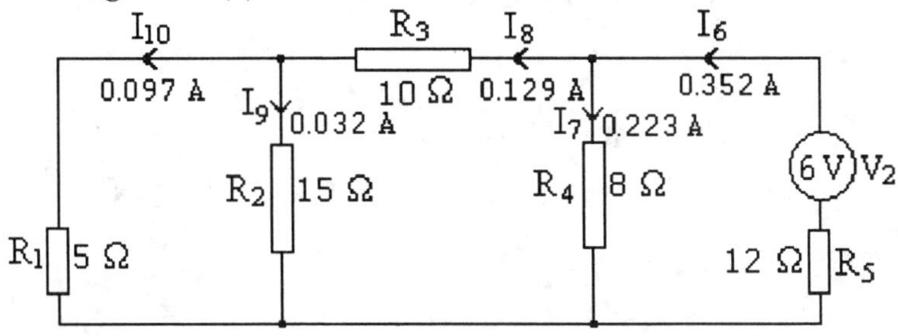

Fig. 2.9-1(e)

65

(contd)

<u>Refer to Fig. 2.9-1(e)</u>

$$I_9 = I_8 \times {}^{R_1}/(R_1 + R_2)$$
$$= 0.129 \times {}^5/(5 + 15) \qquad = \underline{\textbf{0.032 A}}$$
$$I_{10} = I_8 - I_9 = 0.129 - 0.032 = \underline{\textbf{0.097 A}}$$

Finally, superimpose the networks of Fig. 2.9-0(e) on Fig. 2.9-1(e) and determine the current in each branch of the network with both sources of supply as shown in Fig. 2.9-2.

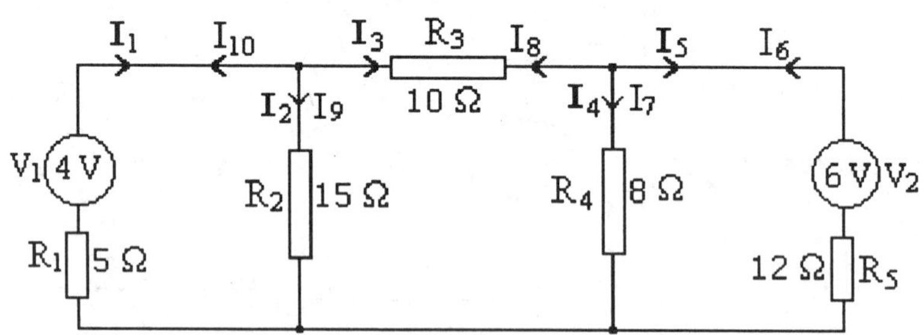

Fig. 2.9-2. Distribution of current by each source

<u>Refer to Fig. 2.9-2</u>

Superimposing the networks, we get

Current in V_1 (R_1) $= I_1 - I_{10}$
$$= 0.321 - 0.097 \qquad = \underline{\textbf{0.224 A}}$$

Current in R_2 (15Ω) $= I_2 + I_9$
$$= 0.159 + 0.032 \qquad = \underline{\textbf{0.191 A}}$$

Current in R_3 (10Ω) $= I_3 - I_8$
$$= 0.162 - 0.129 \qquad = \underline{\textbf{0.033 A}}$$

Current in R_4 (8Ω) $= I_4 + I_7$
$$= 0.0972 + 0.223 \qquad = \underline{\textbf{0.320 A}}$$

Current in V_2 (R_5) $= I_6 - I_5$
$$= 0.352 - 0.0648 \qquad = \underline{\textbf{0.287 A}}$$

Example 2.10

A circuit for charging accumulators has two lamps in parallel, the resistance of one lamp being 2.2 Ω and that of the other 1.3 Ω. Estimate the value of the resistance R in Fig. 2.10-0 that must be connected in series with the accumulators to limit the charging current to 5A. The charging circuit is supplied by a source having an e.m.f of 12 V and internal resistance of 0.12 Ω. The accumulators have a resistance of 0.4 Ω and an e.m.f of 6V.

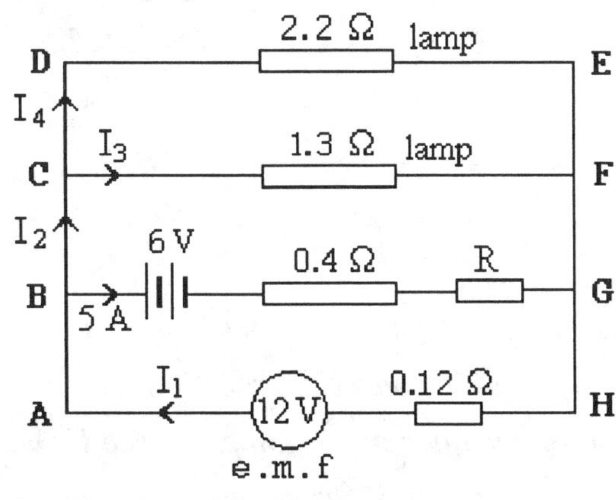

Fig. 2.10-0

Solution

Applying Kirchhoff's Law (voltage law) to Fig. 2.10-0, we get

Closed loop ABGHA:

$$5(0.4 + R) + 0.12\,I_1 = 12 - 6$$
$$2.0 + 5R + 0.12\,I_1 = 6$$
$$5R + 0.12\,I_1 = 4 \quad \text{(Eq. 2.10-0)}$$

Closed Loop ACFHA:

$$1.3\,I_3 + 0.12\,I_1 = 12 \quad \text{(Eq. 2.10-1)}$$

Subtract Eq. 2.10-1 from Eq. 2.10-0

$$5R + 0.12\,I_1 = 4$$
$$\underline{1.3\,I_3 + 0.12\,I_1 = 12}$$
$$-1.3\,I_3 + 5R = -8$$

That is,
$$1.3\,I_3 - 5R = 8 \quad \text{(Eq. 2.10-2)}$$

Closed loop ADEHA: $2.2\,I_4 + 0.12\,I_1 = 12$

(Note $I_4 = I_2 - I_3$ $2.2(I_2 - I_3) + 0.12\,I_1 = 12$

and $I_2 = I_1 - 5$) $2.2[(I_1 - 5) - I_3] + 0.12\,I_1 = 12$

$$2.2\,I_1 - 11 - 2.2\,I_3 + 0.12\,I_1 = 12$$

That is, $2.32\,I_1 - 2.2\,I_3 = 23 \quad \text{(Eq. 2.10-3)}$

(contd)

Subtract Eq. 2.10-0 from Eq. 2.10-3

(Eq. 2.10-3) x 0.12	$0.2784\,I_1 - 0.264\,I_3$	$= 2.76$
(Eq. 2.10-0) x 2.32	$\underline{0.2784\,I_1 \qquad\qquad +\,11.6\,R}$	$= \underline{9.28}$
	$-0.264\,I_3 - 11.6\,R$	$= -6.52$

That is, $\qquad 0.264\,I_3 + 11.6\,R = 6.52$ (Eq. 2.10-4)

Subtract Eq. 2.10-4 from Eq. 2.10-2

(Eq. 2.10-2) x 0.264	$0.3432\,I_3 - 1.32\,R$	$= 2.112$
(Eq. 2.10-4) x 1.3	$\underline{0.3432\,I_3 + 15.08\,R}$	$= \underline{8.476}$
	$-16.4\,R$	$= -6.364$

$$\therefore \quad R = \frac{6.364}{16.4} = \underline{\mathbf{0.388\ \Omega}}$$

Example 2.11

A load of resistance 6 ohms is supplied by a parallel combination of two sources, E_1 and E_2. Source E_1 has an open-circuit e.m.f of 42V and an internal resistance of 12 ohms. Source E_2 has an open-circuit e.m.f of 35V and an internal resistance of 3 ohms. Using the superposition theorem, calculate the current supplied by the 35V source and the power dissipated internally. Refer to Fig. 2.11-0.

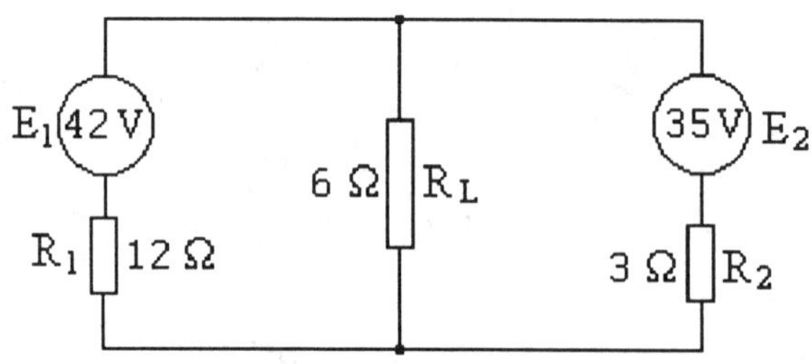

Fig. 2.11-0

Solution

Source E_1 acting alone [Refer to Fig. 2.11-0(a)]

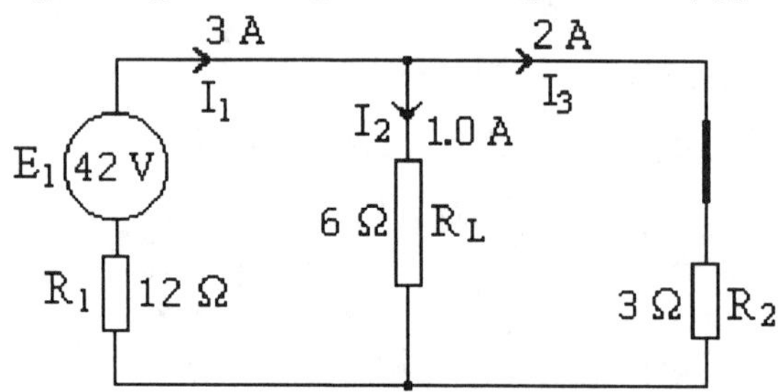

Fig. 2.11-0(a)

(contd)

<u>Refer to Fig. 2.11-0(a)</u>

$$I_1 = \frac{42}{\dfrac{6 \times 3}{6 + 3} + 12} = \underline{3\,A}$$

$$I_2 = 3 \times \frac{3}{6 + 3} = \underline{1.0\,A}$$

$$I_3 = (3 - 1.0)\,A = \underline{2\,A}$$

[Hint: It would be incorrect to calculate the power dissipated in the 3 ohm resistor by the current from E_1 and superimpose with the power dissipated by the current from E_2. Superposition cannot be applied in power calculations, the reason being that power is not linearly related to voltage or current. Power $P = V^2/R$ or I^2R.]

Source E_2 acting alone [<u>Refer to Fig. 2.11-0(b)</u>]

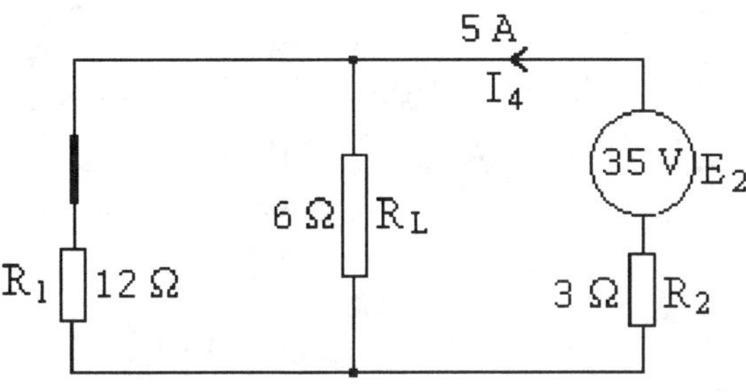

Fig. 2.11-0(b)

Since it is required to determine the current and power supplied and dissipated by E_2, it is only necessary to calculate the current I_4.

$$I_4 = \frac{E_2}{R_2 + \dfrac{R_1\,R_L}{R_1 + R_L}}$$

$$= \frac{35}{3 + \dfrac{12 \times 6}{12 + 6}} = 5\,A$$

Superimpose Fig. 2.11-0(a) on Fig. 2.11-0(b) and calculate the current in E_2.

Current through E_2 $= I_4 - I_3$

$= (5 - 2)\,A \qquad = \underline{\textbf{3 A}}$

Power dissipated in E_2 $= I^2\,R_2$

$= 3^2 \times 3 \qquad = \underline{\textbf{27 W}}$

Example 2.12

*A generator in a motor vehicle has an e.m.f of 16.3 V and an internal
resistance of 0.2 Ω. It is connected in parallel with a battery having an
e.m.f of 12 V and an internal resistance of 0.1 Ω. The combination
supplies the electrical equipment of the vehicle, which is considered
equivalent to a resistance of 4 Ω. Calculate (a) the current from the
generator, (b) the current flowing through the battery, and (c) the p.d
at the battery terminals.*

Solution

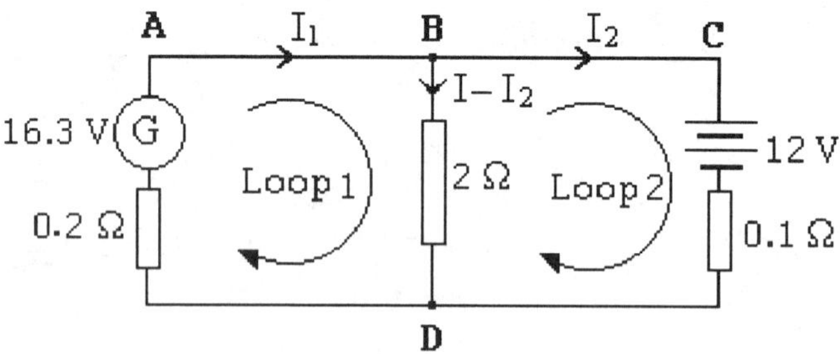

Fig. 2.12-0

<u>Refer to Fig. 2.12-0</u>

Applying Kirchhoff's Law to the respective loops, we get

Loop 1, **ABDA:**
$$0.2\,I_1 + 2\,(I_1 - I_2) = 16.3$$
$$0.2\,I_1 + 2\,I_1 - 2\,I_2 = 16.3$$
$$2.2\,I_1 - 2\,I_2 = 16.3 \qquad \text{(Eq. 2.12-0)}$$

Loop 2, **BCDB:**
$$0.1\,I_2 - 2\,(I_1 - I_2) = 12$$
$$0.1\,I_2 - 2\,I_1 + 2\,I_2 = 12$$
$$- 2\,I_1 + 2.1\,I_2 = 12 \qquad \text{(Eq. 2.12-1)}$$

Subtracting Eq. 2.12-1 from Eq. 2.12-0, we get

(Eq. 2.12-0) x 2 $\quad 4.4\,I_1 - 4\,I_2 = 32.6$

(Eq. 2.12-1) x 2.2 $\quad \underline{4.4\,I_1 + 4.62\,I_2 \qquad = 26.4}$

$$\qquad\qquad\qquad\qquad - 0.62\,I_2 = 6.2$$

$$\therefore \qquad I_2 = -\left(\frac{6.2}{0.62}\right) = -\underline{\mathbf{10\ A}}$$

[The minus sign indicates that I_2 flows in the reverse direction to
that indicated by the arrowhead in Fig. 2.12-0.]

70

(contd)

Substitute for I_2 in Eq. 2.12-0 and hence determine the value of the current I_1 coming from the generator (G).

$$\text{Thus} \qquad 2.2\,I_1 - 2\,(-10) = 16.3$$
$$2.2\,I_1 = 36.3$$
$$\therefore \qquad I_1 = 36.3/2.2 = \underline{16.5\ \text{A}}$$

(a)   **Current from Generator**   = <u>**16.5 A**</u>

(b)   **Current in battery**   = <u>**10 A**</u>

(c)   **P.d at battery terminals** = $12 + (10 \times 0.1)$ = <u>**13 V**</u>

Example 2.13

Determine the resistance between the points Y and Z of the circuit shown in Fig. 2.13-0 using the delta-star transformation.

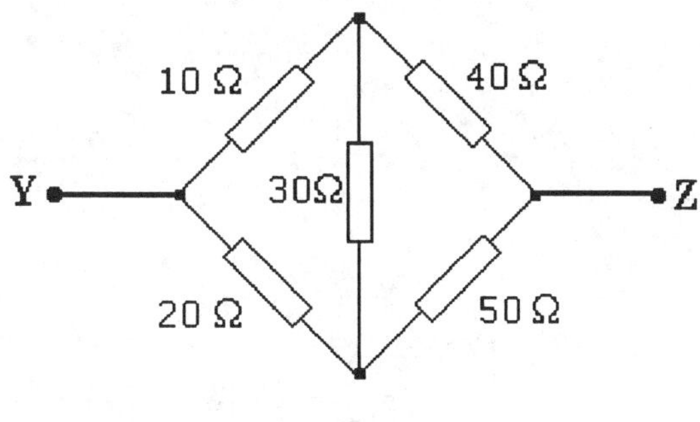

Fig. 2.13-0

Solution

Redraw Fig. 2.13-0 and label the nodes as shown in Fig. 2.13-0(a)

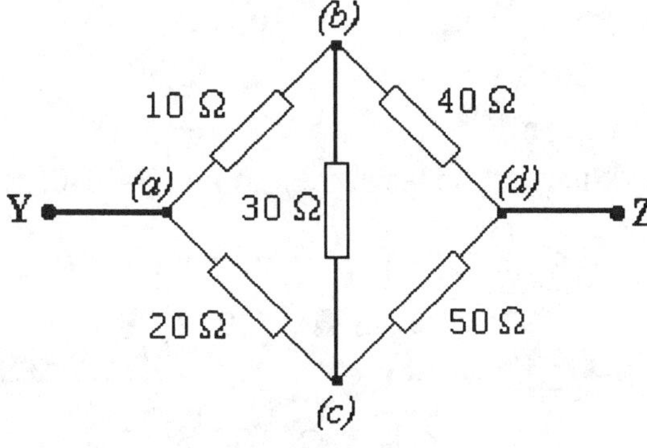

Fig. 2.13-0(a)

(contd)

<u>Refer to Fig. 2.13-0(a)</u>

Convert the delta-connected resistors between points *(a), (b) and (c)* to a star network.

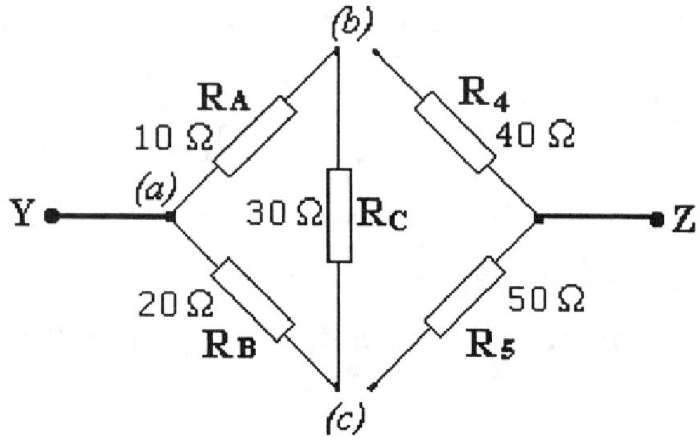

Fig. 2.13-0(b)

The conversion of the delta circuit R_A, R_B and R_C of Fig. 2.13-0(b) formed by nodes *(a), (b) and (c)* to a star network is shown in Fig. 2.13-0(c).

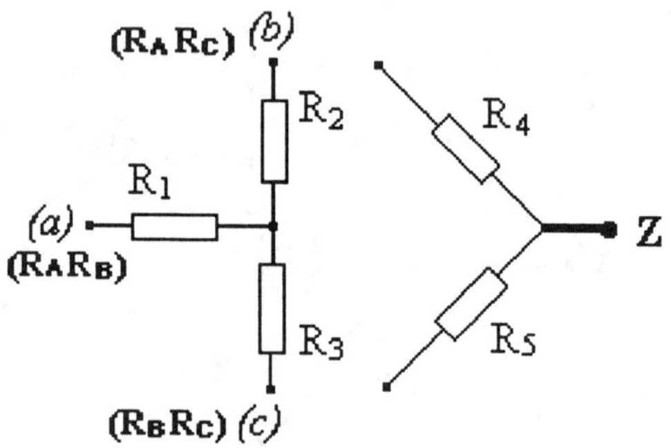

Fig. 2.13-0(c)

<u>Refer to Fig. 2.13-0(c)</u>

$$R_1 = \frac{R_A \, R_B}{R_A + R_B + R_C} = \frac{10 \times 20}{10 + 20 + 30} = \frac{200}{60} = \underline{3.33\ \Omega}$$

$$R_2 = \frac{R_A \, R_C}{R_A + R_B + R_C} = \frac{10 \times 30}{10 + 20 + 30} = \frac{300}{60} = \underline{5\ \Omega}$$

$$R_3 = \frac{R_B \, R_C}{R_A + R_B + R_C} = \frac{20 \times 30}{10 + 20 + 30} = \frac{600}{60} = \underline{10\ \Omega}$$

The network is redrawn as shown in Fig. 2.13-0(d).

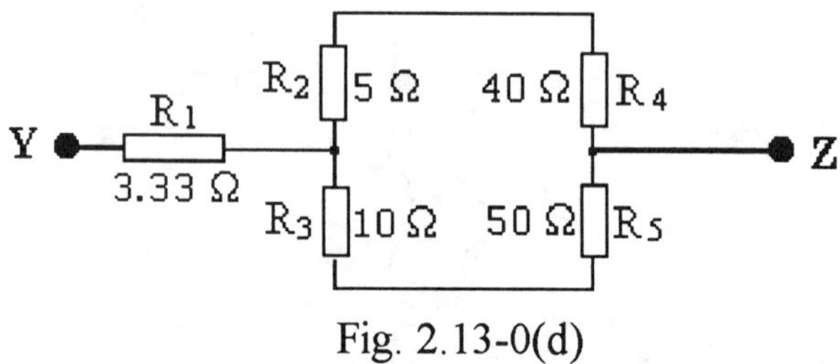

Fig. 2.13-0(d)

(contd)

<u>Refer to Fig. 2.13-0(d)</u>

From inspection, $(R_2 + R_4)$ is in parallel with $(R_3 + R_5)$.

$$\text{Resistance between } \mathbf{Y}\,\mathbf{Z} = R_1 + \frac{(R_2 + R_4)\,(R_3 + R_5)}{R_1 + R_2 + R_3 + R_5}$$

$$= 3.33 + \frac{(5 + 40)\,(10 + 50)}{5 + 40 + 10 + 50}$$

$$= 3.33 + \frac{(45)\,(60)}{105}$$

$$= 3.33 + 25.71 = \underline{\mathbf{29.04\ \Omega}}$$

Example 2.14

Two d.c generators of 120 V and 140 V have internal resistances of 10 Ω and 15 Ω, respectively. The two generators are connected in parallel to feed a lighting and a heating load having an effective load resistance of 10 Ω. Calculate the current delivered to the load and by each generator.

Solution

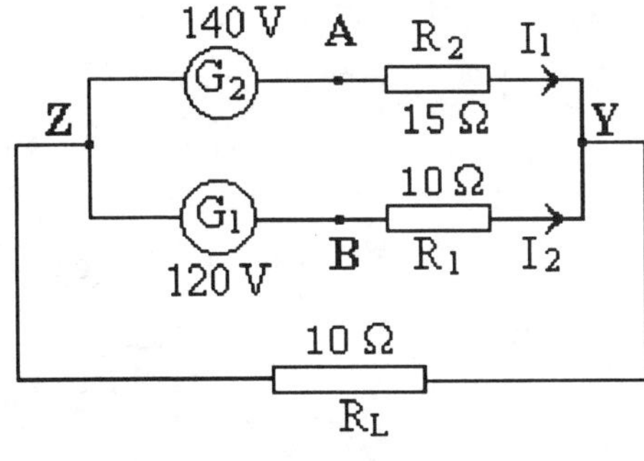

Fig. 2.14-0(a)

<u>Refer to Fig. 2.14-0(a)</u>.

We will apply Thevenin's Theorem to resolve the current.

Step 1: Remove the load resistor R_L and determine the open-circuit voltage between the terminals **Z Y**.

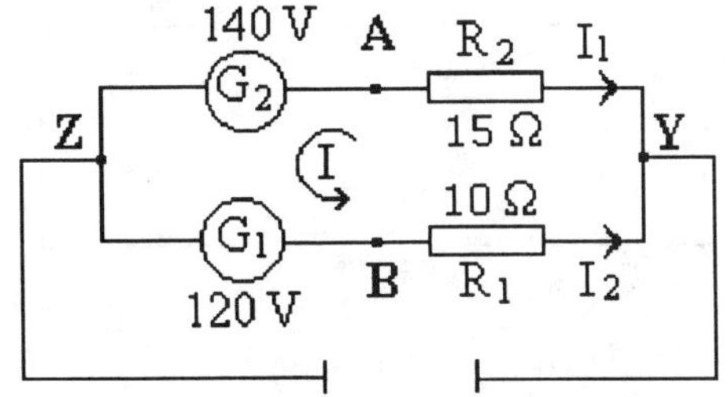

Fig. 2.14-0(b). Load R_L removed

73

(contd)

<u>Refer to Fig. 2.14-0(b)</u>

Circulating current, $I = \dfrac{V_2 - V_1}{R_1 + R_2}$ $(V_2 = G_2$ and $V_1 = G_1)$

$$= \dfrac{140 - 120}{10 + 15} \qquad = \underline{0.8\ A}$$

Voltage between terminals **Z Y** is

$$V_{ZY} = V_2 - (I \times 15)$$
$$= 140 - (0.8 \times 15)$$
$$= 140 - 12 \qquad = \underline{128\ V}$$

OR $\qquad V_{ZY} = V_1 + (I \times 10)$
$$= 120 + (0.8 \times 10) \quad = \underline{128\ V}$$

Step 2: Short-circuit all sources of e.m.f (the generators) and replace them by their internal resistances. Hence, calculate the effective internal resistance **r** of the network. <u>Refer to Fig. 2.14-0(c)</u>.

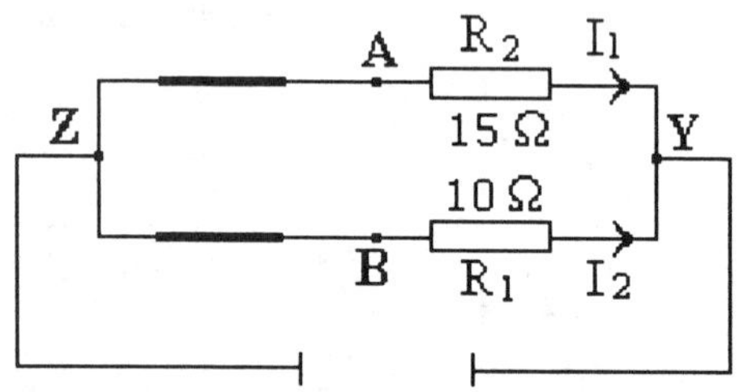

Fig. 2.14-0(c)

Effective resistance, **r** $= \dfrac{R_1 R_2}{R_1 + R_2}$

$$= \dfrac{10 \times 15}{10 + 15} \qquad = \underline{6.0\ \Omega}$$

Step 3: The original network is now represented by a single source of e.m.f (that is, $V_{ZY} = 128$ V) having an internal resistance **r**. Calculate the value of the current I_L in the load resistance, R_L. [Fig. 2.14-0(d)]

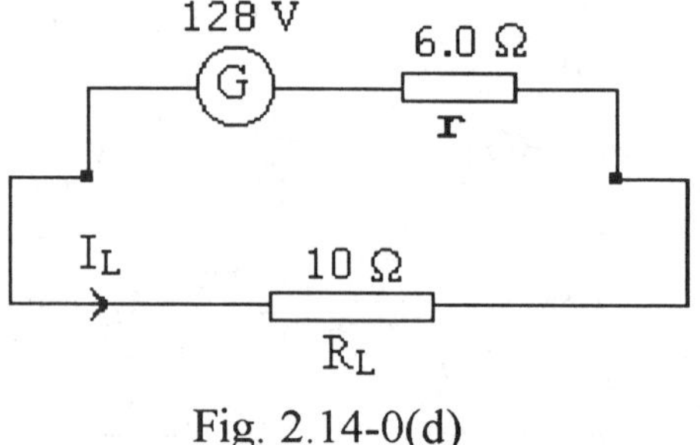

Fig. 2.14-0(d)

(contd)

Refer to Fig. 2.14-0(d)

By Thevenin's Theorem, the current flowing in the load resistor R_L is

$$I_L = \frac{V_{ZY}}{R_L + r} = \frac{128}{10 + 6} = \underline{8\ A}$$

Voltage across R_L $= V_{RL} = I_L \times R_L = 8 \times 10 = \underline{80\ V}$

We can now determine the current supplied by each generator.

Current supplied by $G_2 = I_2$

Now $\quad V_{RL} = V_2 - (15 \times I_2)$

That is $\quad 80 = 140 - 15\ I_2$

$\therefore \quad I_2 = \dfrac{140 - 80}{15} = \underline{\mathbf{4.0\ A}}$

Current supplied by $G_1 = I_1$

$V_{RL} = V_1 - (10 \times I_1)$

$80 = 120 - 10\ I_1$

$\therefore \quad I_1 = \dfrac{120 - 80}{10} = \underline{\mathbf{4.0\ A}}$

Example 2.15

A distributor 2700 m in length has the following loads connected to it at 1800 m, 2200 m, and 2700 m from the distribution point, 65 A, 20 A and 15 A, respectively. The resistance of the distributor (single cable) is 0.18 Ω for the 1800 m section, 0.22 Ω for the 2200 m length and 0.27 Ω for the final length. Calculate (a) the current in each section of the distributor, (b) the power loss in each section, and (c) the supply voltage at the feeding point if the p.d across the 15 A load is 440 V

Solution

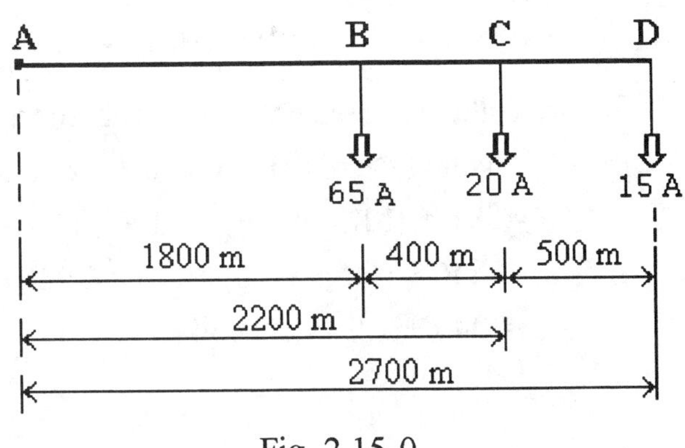

Fig. 2.15-0

(contd)

<u>Refer to Fig. 2.15-0</u>

(a)   Applying Kirchhoff's Laws (current law) to the distributor circuit, we get

Current in section A B   = 65 A + 20 A + 15 A   = **100 A**

Current in section B C   = (100 − 65) A   = **35 A**

Current in section C D   = 100 A − (65 + 20) A   = **15 A**

OR   Current in section C D   = (35 − 20) A   = **15 A**

(b)   The circuit is now redrawn as shown in Fig. 2.15-0(a).

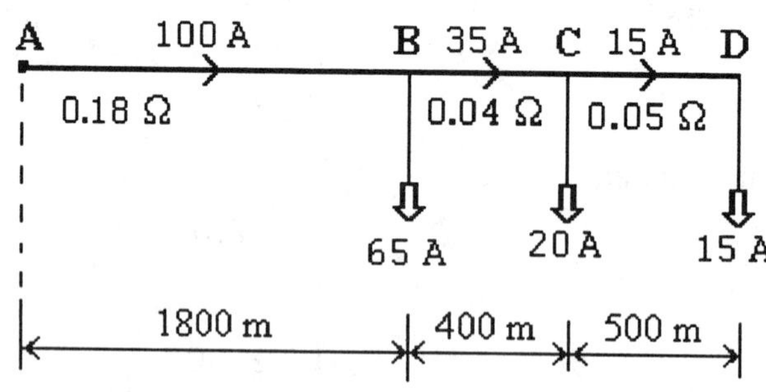

Fig. 2.15-0(a)

Distributor resistances in each section for both cables are as follows:

$$R_{AB} = 2 \times 0.18 \qquad = \underline{0.36\ \Omega}$$

$$R_{BC} = 2(0.22 - 0.18) \qquad = \underline{0.08\ \Omega}$$

$$R_{CD} = 2(0.27 - 0.22) \qquad = \underline{0.10\ \Omega}$$

Power loss in section   A B $= I^2 R_{AB}$

$$= 100^2 \times 0.36 \qquad = \underline{\textbf{3600 W}}$$

Power loss in section   B C $= I^2 R_{BC}$

$$= 35^2 \times 0.08 \qquad = \underline{\textbf{98 W}}$$

Power loss in section   C D $= I^2 R_{CD}$

$$= 15^2 \times 0.10 \qquad = \underline{\textbf{22.5 W}}$$

(c)   Supply voltage at feed point of distributor is

= voltage across 15 A load + voltage drop along distributor

$= V_L + (I\,R_{CD} + I\,R_{BC} + I\,R_{AB})$

$= 440 + [(15 \times 0.10) + (35 \times 0.08) + (100 \times 0.36)]$

$= 440 + 1.5 + 2.8 + 36 \qquad = \underline{\textbf{480.3 V}}$

Example 2.16

A distributor is fed from both ends at 240 V. Loads of 50 A, 60 A and 20 A are connected at points B, C and D. The distributor cable resistance (both cables) from A to B is 0.1 Ω, from B to C is 0.08 Ω, from C to D is 0.09 Ω and from D to E is 0.1 Ω. Determine the distribution of current in the distributor and the p.d available at each of the feeding points B, C and D.

Solution

A single-line diagram is drawn to represent the distributor as shown in Fig. 2.16-0.

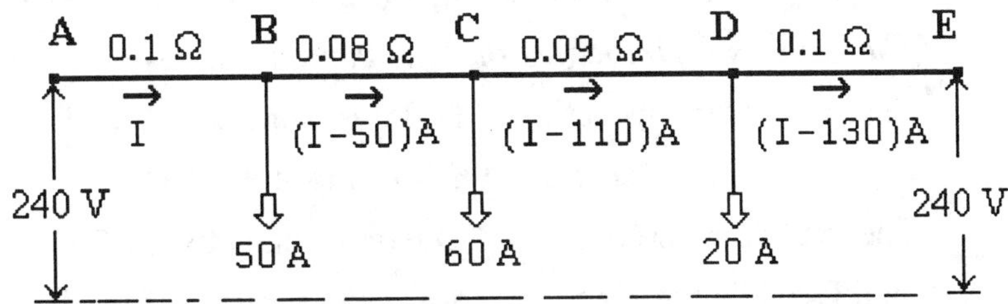

Fig. 2.16-0. Single-line diagram representing distributor

Let us assume that the current I flows from end A. Then by applying Kirchhoff's Laws (current law), we get

$$\text{Current in section B C} = (I - 50) \text{ A}$$
$$\text{Current in section CD} = [I - (50 + 60)] \text{ A}$$
$$= (I - 110) \text{ A}$$
$$\text{Current in section D E} = [I - (50 + 60 + 20)]$$
$$= (I - 130) \text{ A}$$

<u>Refer to Fig. 2.16-0(a)</u>. The distributor is fed from both ends at 240V. Therefore, applying Kirchhoff's Laws (voltage law), we get

$$(0.1 \times I) + 0.08 (I - 50) + 0.09 (I - 110) + 0.1 (I - 130) = 0$$
$$0.1 I + 0.08 I - 4 + 0.09 I - 9.9 + 0.1 I - 13 = 0$$
$$0.37 I - 26.9 = 0$$
$$\therefore \quad I = {}^{26.9}/_{0.37}$$
$$= \underline{\mathbf{72.7 \text{ A}}}$$

Substitute the value for I (72.7 A) as shown in Fig. 2.16-0(a) and then determine the current in each section of the distributor.

(contd)

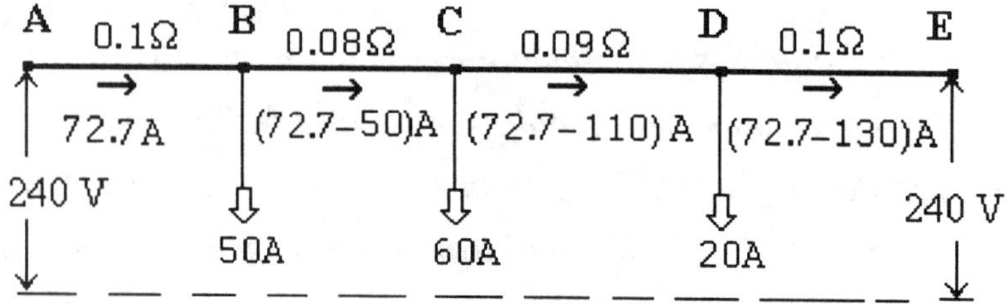

Fig. 2.16-0(a). Single-line diagram representing distributor

<u>Refer to Fig. 2.16-0(a)</u>

Current in section A B   = **72.7 A**   (that is, I = 72.7 A)

Current in section B C   = (72.7 − 50) A   =   **22.7 A**

Current in section C D   = (72.7 − 110) A   = − **37.3 A**

Current in section D E   = (72.7 − 130) A   = − **57.3 A**

The minus sign indicates that the current flow is in the opposite direction.

That is, current in section C D flows from D to C and

      current in section D E flows from E to D.

The load connected at point D is fed totally by the 240 V supply from

end E, while the load at point C shares current from both ends.

The final distribution is illustrated in Fig. 2.16-0(b).

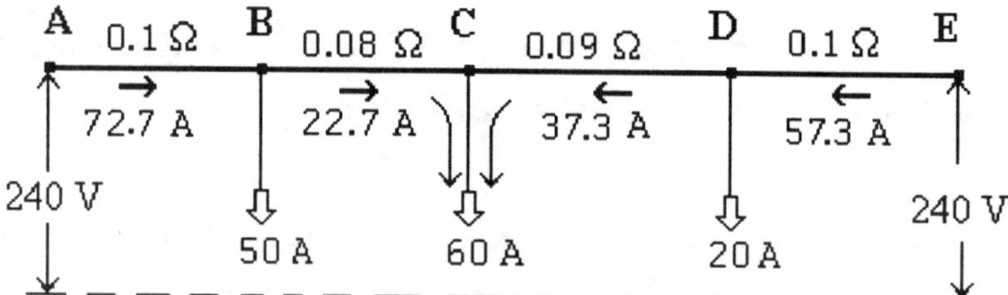

Fig. 2.16-0(b). Single line diagram representing distributor

<u>Refer to Fig. 2.16-0(b)</u>

Observe that load C = 60 A receives current from both ends, that is

      from end A   = 22.7 A

and   from end E   = <u>37.3 A</u>

        Total   = <u>60.0 A</u>

The p.d at each feeding point can now be determined as follows:

$$\text{P.d at point B} = 240 - (I \times R_{AB})$$

$$= 240 - (72.7 \times 0.1)$$

$$= 240 - 7.27 \qquad = \underline{\textbf{232.73 V}}$$

$$\text{P.d at point C} = 232.73 - (22.7 \times 0.08)$$

$$= 232.73 - 1.82 \qquad = \underline{\textbf{230.91 V}}$$

(contd)

$$\text{P.d at point D} = 230.91 - (-37.3 \times 0.09)$$
$$= 230.91 + 3.36 \qquad = \underline{\mathbf{234.27\ V}}$$

P.d at end E: This computation is not required because end E is at 240 V. However, being curious, let us compute the voltage.

$$= 234.27 - (-57.3 \times 0.1)$$
$$= 234.27 + 5.73 \qquad = \underline{\mathbf{240\ V}}$$

Example 2.17

A distributor is fed at both ends A and B. At end A the voltage is maintained at 236 V and at end B at 235 V. The length of the distributor is 200 yd and loads are tapped off from end A as follows:

20 A at 50 yd

40 A at 75 yd

25 A at 100 yd

30 A at 150 yd

The resistance per 1000 yd of one conductor is 0.4 Ω. Calculate

(a)   the currents in the various sections of the distributor

(b)   the minimum voltage and

(c)   the point at which it occurs

Solution

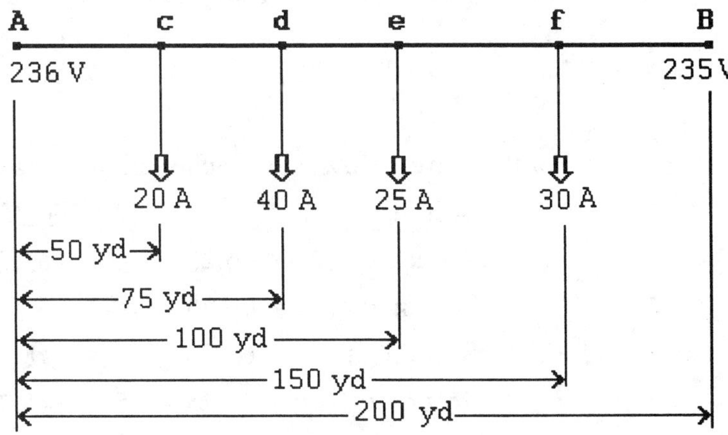

Fig. 2.17-0

Refer to Fig. 2.17-0

Resistance of 50 yd of distributor (both conductors) is

$$R_{AC} = 2 \left(\frac{0.4 \times 50}{1000} \right) = 0.04\ \Omega$$

(contd)

$$\text{Length of section c d} = 25 \text{ yd}$$

$$\therefore \quad R_{cd} = 2 \frac{(0.4 \times 25)}{1000} = \underline{0.02\ \Omega}$$

$$\text{Length of section d e} = 25 \text{ yd}$$

$$R_{de} = \underline{0.02\ \Omega}$$

$$\text{Length of section e f} = 50 \text{ yd}$$

$$R_{ef} = \underline{0.04\ \Omega}$$

$$\text{Length of section f B} = 50 \text{ yd}$$

$$R_{fB} = \underline{0.04\ \Omega}$$

Let the current from end A be I amperes. Then the current in each section is as shown in Fig. 2.17-0(a).

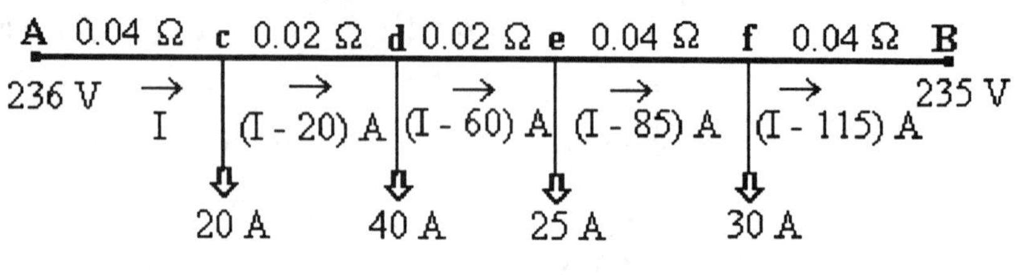

Fig. 2.17-0(a)

<u>Refer to Fig. 2.17-0(a)</u>

Applying Kirchhoff's Laws (current law) to path A c d e f B, we get

$$0.04\,I + 0.02(I - 20) + 0.02(I - 60) + 0.04(I - 85) + 0.04(I - 115)\ldots$$

$$= (236 - 235) \text{ V}$$

$$0.04\,I + 0.02\,I - 0.4 + 0.02\,I - 1.2 + 0.04\,I - 3.4 + 0.04\,I - 4.6 = 1.0$$

$$0.16\,I - 9.6 = 1.0$$

$$\therefore \quad I = \frac{1 + 9.6}{0.16} = \underline{\mathbf{66.25\ A}}$$

(a)   For the current flowing in each section, we get

$$\text{Section A c} = I = \underline{\mathbf{66.25\ A}}$$

$$\text{Section c d} = 66.25 - 20 = \underline{\mathbf{46.25\ A}}$$

$$\text{Section d e} = 66.25 - 60 = \underline{\mathbf{6.25\ A}}$$

$$\text{Section e f} = 66.25 - 85 = -\underline{\mathbf{18.75\ A}}$$

$$\text{Section f B} = 66.25 - 115 = -\underline{\mathbf{48.75\ A}}$$

The current in sections **e f** (– 18.75 A) and **f B** (– 48.75 A) flows in the opposite direction initially indicated.

The correct direction and value of current in each section of the distributor are now shown in Fig. 2.17-0(b).

(contd)

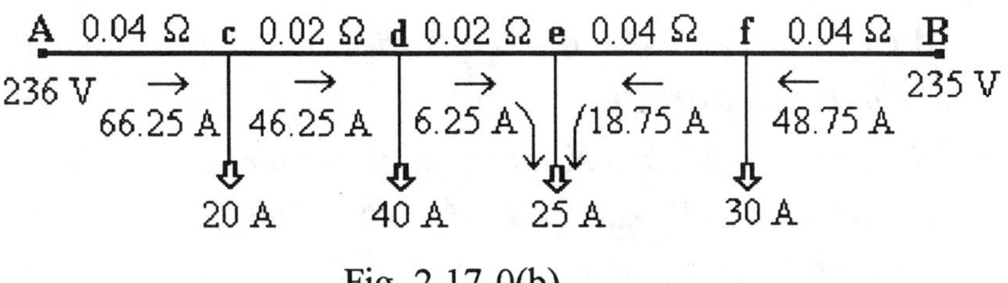

Fig. 2.17-0(b)

Observe that the 25A load current is subscribed from both ends.

(b) The p.d at each load point is as follows:

$$\text{At 20 A load} = 236 - (66.25 \times 0.04)$$
$$= 236 - 2.65 \qquad = \underline{\textbf{233.35 V}}$$

$$\text{At 40 A load} = 233.35 - (46.25 \times 0.02)$$
$$= 233.35 - 0.925 \qquad = \underline{\textbf{232.425 V}}$$

$$\text{At 25 A load} = 232.425 - (6.25 \times 0.02)$$
$$= 232.425 - 0.125 \qquad = \underline{\textbf{232.3 V}}$$

Working from supply end B , we get

$$\text{P.d at 30 A load} = 235 - (48.75 \times 0.04)$$
$$= 235 - 1.95 \qquad = \underline{\textbf{233.05 V}}$$

We will continue to work from the supply end B.

The p.d at the 25 A load point was calculated from the supply end A. However, we will also compute it from end B, which should give the same result. Thus

$$\text{P.d at 25 A load} = 233.05 - (18.75 \times 0.04)$$
$$= 233.05 - 0.75 \qquad = \underline{\textbf{232.3V}}$$

(c) The minimum p.d $= \underline{\textbf{232.3 V}}$

This occurs at the 25 A load point 100 yd from supply end A, or 100 yd from supply end B.

Example 2.18

A three-wire, (+ 240 – 0 – 240) V distribution system supplies the
following loads:
Between + 240 V and N, a load of 60 A, 50 m away
Between + 240 V and N, a load of 40 A, 100 m away
Between – 240 V and N, a load of 30 A, 75 m away
The resistance of each distributor cable is 0.8 Ω per 1000 m and the
neutral conductor is half the cross-sectional area of the outers.
Calculate the p.d across each load.

Solution

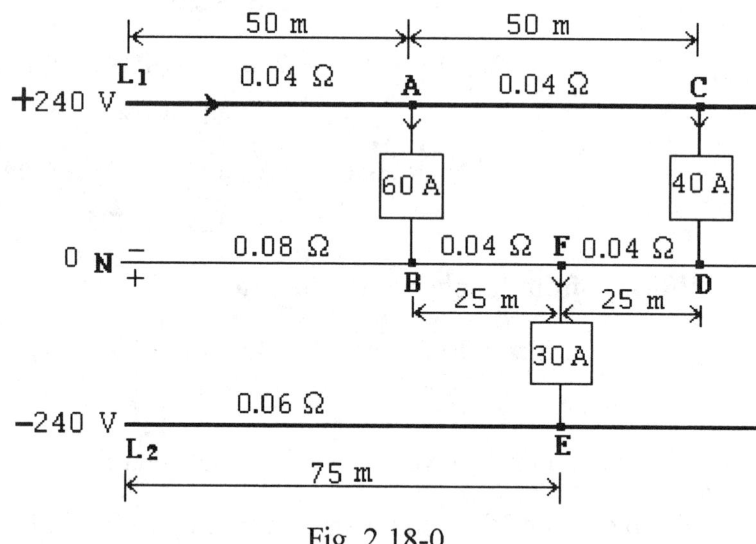

Fig. 2.18-0

Refer to Fig. 2.18-0

Resistance of section L₁A $\quad = \dfrac{0.8 \times 50}{1000} \quad = 0.04\ \Omega$

Resistance of section A C $\quad = 0.04\ \Omega$

Resistance of section L₂E $\quad = \dfrac{0.8 \times 75}{1000} \quad = 0.06\ \Omega$

Since the cross-sectional area of the neutral conductor is half that of the
area of the outers, its resistance is 2 x L₁ or 2 x L₂.

$\therefore$ Resistance of section N B $\; = 2 \times 0.04 \qquad = 0.08\ \Omega$

Resistance of section B F = F D $\; = 2\left(\dfrac{0.8 \times 25}{1000}\right) \qquad = 0.04\ \Omega$

The distribution system is redrawn in Fig. 2.18-0(a) showing current
distributions and their directions in each section.

(contd)

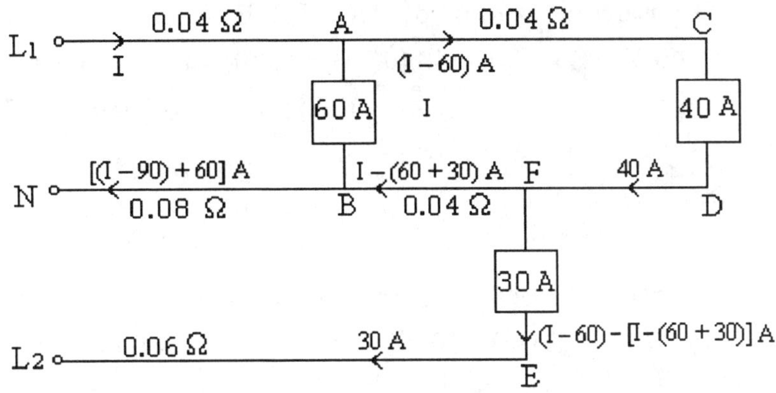

Fig. 2.18-0(a)

<u>Refer to Fig. 2.18-0(a)</u>

Current in section $L_1A = I\ = 60 + 40\qquad\qquad = \underline{100\ A}$

Current in section A C $= I - 60\ = 100 - 60\qquad = \underline{40\ A}$

Current in section F B $= I - (60 + 30)\ = 100 - 90\ = \underline{10\ A}$

Current in section B N $= (I - 90) + 60\ = 160 - 90 = \underline{70\ A}$

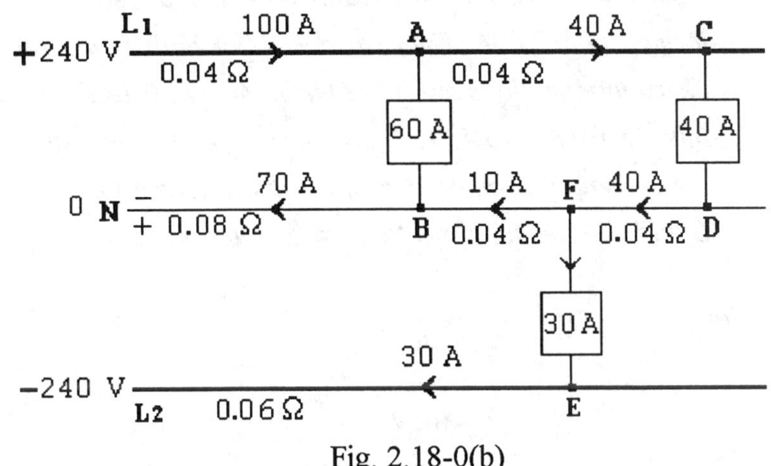

Fig. 2.18-0(b)

<u>Refer to Fig. 2.18-0(b)</u>

The actual currents and their directions of flow are shown in each conductor and section of the distribution system.

Consider the distribution path $L_1A\ B\ N\ L_1$.

By the application of Kirchhoff's Laws (voltage law), we get

$$100\ (0.04) + V_{AB} + 70\ (0.08)\ = 240\ V$$
$$4 + V_{AB} + 5.6\ = 240\ V$$
$$V_{AB}\ = 240 - (5.6 + 4)$$
$$= 240 - 9.6$$
$$\mathbf{V_{AB}\ = \underline{\mathbf{230.4\ V}}}$$

(contd)

Consider distribution path A C D F B A.

$$40 (0.04) + V_{CD} + 40 (0.04) + 10 (0.04) - 230.4 \quad = 0 \quad V$$
$$1.6 + V_{CD} + 1.6 + 0.4 \quad = 230.4 \ V$$
$$V_{CD} + 3.6 \quad = 230.4 \ V$$
$$V_{CD} \quad = 230.4 - 3.6 \ V$$
$$V_{CD} \quad = \underline{\mathbf{226.8 \ V}}$$

Consider distribution path N B F E L₂ N.

$$(- 70 \times 0.08) + (- 10 \times 0.04) + V_{FE} + (30 \times 0.06) \quad = 240 \ V$$
$$- 5.6 - 0.4 + V_{FE} + 1.8 \quad = 240 \ V$$
$$V_{FE} + 1.8 - (5.6 + 0.4) \quad = 240 \ V$$
$$V_{FE} \quad = 240 + 4.2 \ V$$
$$V_{FE} \quad = \underline{\mathbf{244.2 \ V}}$$

Example 2.19

*A substation has its distribution system arranged in a closed delta
formation A B C A. The resistances (for both conductors) of the respective
distributor sections are 0.2 Ω from A to B, 0.15 Ω from B to C and
0.05 Ω from C to A. The loads are 60 A at B and 40 A at C. Calculate
the current in each section of the distributor and the available p.d at
B and C if the supply voltage at A is 480 V.*

Solution

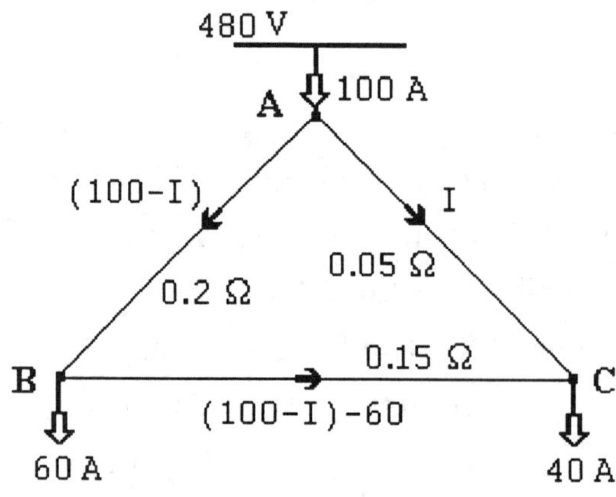

Fig. 2.19-0

(contd)

<u>Refer to Fig. 2.19-0</u>

Consider the loop A B C A:

$$0.2 (100 - I) + 0.15 [(100 - I) - 60] - 0.05 I \quad = 0$$
$$20 - 0.2 I + [(15 - 0.15 I) - 9.0] - 0.05 I \quad = 0$$
$$- 0.4 I + 26 \quad = 0$$
$$I \quad = 26/0.4 = \underline{\mathbf{65\ A}}$$

$$\text{Let}\quad \text{current in section A C} = I \quad = \underline{\mathbf{65\ A}}$$
$$\text{then}\quad \text{current in section A B} = 100 - 65 \quad = \underline{\mathbf{35A}}$$
$$\text{and}\quad \text{current in section B C} = 40 - 65 \quad = \underline{\mathbf{-25\ A}}$$

The current in section B C is -25 A. The negative sign means that the current in section B C flows in the opposite direction to that which is indicated; that is, 25 A flows from C to B (C $\rightarrow$ B).

$$\text{P.d at B} \quad = 480 - (0.2 \times 35) \quad = \underline{\mathbf{473\ V}}$$

$$\text{P.d at C} \quad = 480 - (0.05 \times 65) \quad = \underline{\mathbf{476.75\ V}}.$$

Example 2.20

For the distribution network shown in Fig. 2.20-0 determine

    (a)   the voltage across the load R_L

    (b)   the direction and magnitude of current in the load R_L

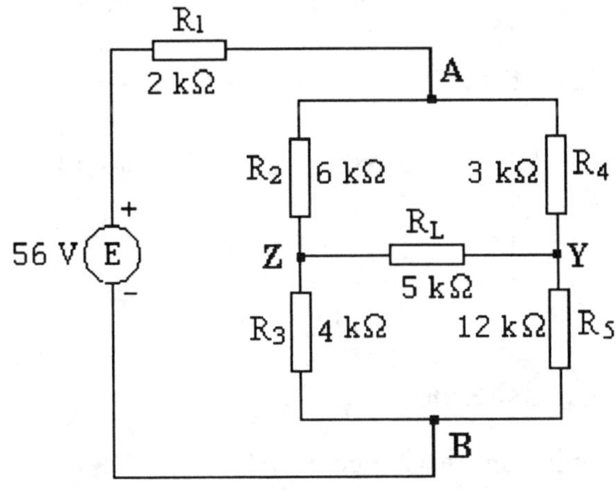

Fig. 2.20-0

(contd)

Solution

The solution of this network can be accomplished by using Kirchhoff's Laws, Thevenin's Theorem, Norton's Theorem, the delta-star transformation or a combination.

Our procedures will take the following directions:

(1)   Remove the load R_L and obtain the open-circuit voltage at the junction of R_2 R_3 and R_4 R_5, that is, the voltage at Z Y.

(2)   With the load still removed, short-circuit Z Y and determine the short-circuit current and its direction.

(3)   Calculate the resistance of the (internal resistance to the open-circuit voltage) circuit using the open-circuit voltage and short-circuit current.

(4)   Draw the equivalent network with the open-circuit voltage [obtained in (1) above] as the source having an internal resistance (r) [obtained in (3) above] in series with R_L.

(5)   Finally determine the voltage across the load R_L.

Step 1:   Load R_L is removed as shown in Fig. 2.20-0(a)

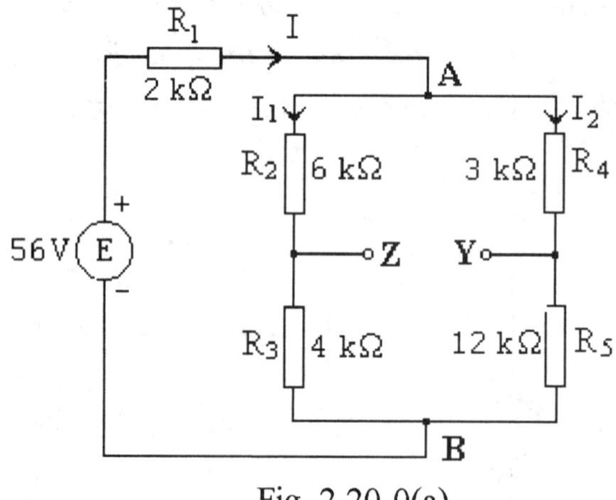

Fig. 2.20-0(a)

<u>Refer to Fig. 2.20-0(a)</u>

Equivalent resistance of **AZB** and **AYB** is

$$R_{eq\,1} = \frac{(R_2 + R_3)\,(R_4 + R_5)}{(R_2 + R_3) + (R_4 + R_5)}$$

$$= \frac{10 \times 15}{10 + 15} = \frac{150}{25} = \underline{6\ k\Omega}$$

(contd)

$$\text{Total circuit resistance } R_T = R_1 + R_{eq\,1} = 2 + 6 \qquad = \underline{8\ k\Omega}$$

$$\text{Current } I = E/R_T = {}^{56}/_8 \qquad = \underline{7\ mA}$$

$$\text{Current } I_1 = \frac{I \times (R_4 + R_5)}{(R_2 + R_3) + (R_4 + R_5)} = \frac{7 \times 15}{25} \qquad = \underline{4.2\ mA}$$

$$\text{Voltage drop in } R_2 = I_1 \times R_2 = 4.2 \times 6 \qquad = \underline{25.2\ V}$$

$$\text{Voltage across } R_3 = I_1 \times R_3 = 4.2 \times 4 \qquad = \underline{16.8\ V}$$

$$\text{Hence, } \mathbf{V_Z} = \underline{16.8\ V \text{ (measured with respect to B)}}$$

$$\text{Current } I_2 = \frac{I \times (R_2 + R_3)}{(R_2 + R_3) + (R_4 + R_5)}$$

$$= \frac{7 \times 10}{25} \qquad = \underline{2.8\ mA}$$

$$\mathbf{OR} \quad I_2 = I - I_1 = 7 - 4.2 \qquad = \underline{2.8\ mA}$$

$$\text{Voltage drop in } R_4 = I_2 \times R_4 = 2.8 \times 3 \qquad = \underline{8.4\ V}$$

$$\text{Voltage across } R_5 = I_2 \times R_5 = 2.8 \times 12 \qquad = \underline{33.6\ V}$$

$$\text{Hence, } \mathbf{V_Y} = \underline{33.6\ V \text{ (measured with respect to B)}}$$

Alternatively

$$\text{Voltage across, } \mathbf{A\,B} = E - I\,R_1$$

$$\text{That is, } V_{AB} = 56 - (7 \times 2) = \underline{42\ V}$$

$$I_1 = \frac{V_{AB}}{R_2 + R_3} = \frac{42}{10} \qquad = \underline{4.2\ mA}$$

$$V_Z = I_1 R_3 = 4.2 \times 4 \qquad = \underline{16.8\ V}$$

$$I_2 = \frac{V_{AB}}{R_4 + R_5} = \frac{42}{15} \qquad = \underline{2.8\ mA}$$

$$V_Y = I_2 \times R_5 = 2.8 \times 12 \qquad = \underline{33.6\ V}$$

$V_Y = 33.6$ V is more positive with respect V_Z; therefore current will flow from **Y** to **Z**

Step 2: With load still removed, short circuit junction **ZY**, then determine the short-circuit current.

Refer to Fig. 2.20-0(b)

Equivalent resistance of parallel branches $R_2\,R_4$ and $R_3\,R_5$ is

$$R_{eq\,2} = \frac{R_2\,R_4}{R_2 + R_4} + \frac{R_3\,R_5}{R_3 + R_5}$$

$$= \frac{6 \times 3}{9} + \frac{4 \times 12}{16} = \underline{5\ k\Omega}$$

(contd)

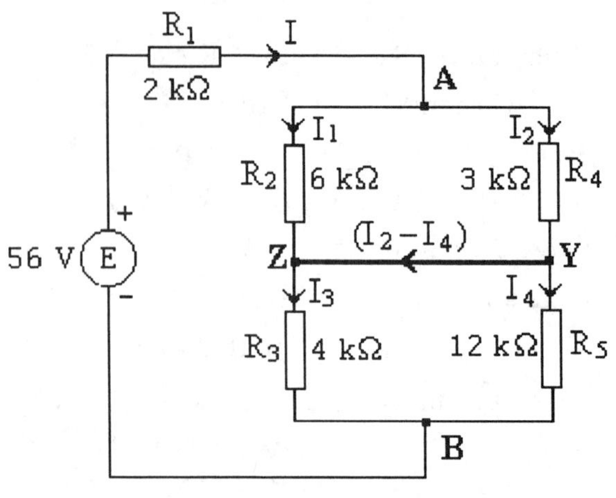

Fig. 2.20-0(b)

$$\text{Total circuit resistance} = R_1 + R_{eq\,2} = 2 + 5 = \underline{7\,k\Omega}$$
$$\text{Current} \quad I = E/R_T = {}^{56}/7 = \underline{8\,mA}$$

By inspection of Fig. 2.20-0(b), current I_2 is greater than I_1. Therefore direction of short-circuit current flow is from Y to Z (Y $\rightarrow$ Z).

$$\text{Current } I_2 \text{ through } R_4 = \frac{I \times R_2}{R_2 + R_4} = \frac{8 \times 6}{9} = \underline{5.33\,mA}$$

$$\text{Current } I_4 \text{ through } R_5 = \frac{I \times R_3}{R_3 + R_5} = \frac{8 \times 4}{16} = \underline{2.0\,mA}$$

$$\text{Short-circuit } I_{YZ} = (I_2 - I_4) = 5.33 - 2.0 = \underline{3.33\,mA}$$

$$\text{Direction of current is } \mathbf{Y \rightarrow Z}$$

Step 3: The resistance r of the circuit (or internal resistance r to the open-circuit voltage) is

$$r = \frac{V_{YZ}}{I_{YZ}} = \frac{16.8}{3.33} = \underline{5.045\,k\Omega}$$

Step 4: The equivalent network is drawn with open-circuit voltage as the source and r as its internal resistance in series with R_L.

<u>Refer to Fig. 2.22-0(c)</u>

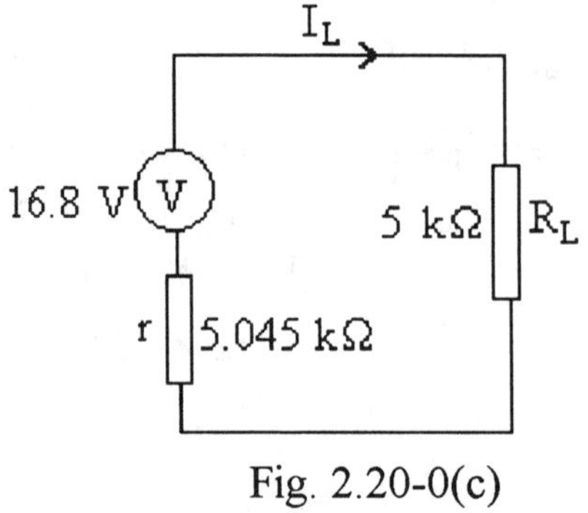

Fig. 2.20-0(c)

(contd)

Step 5: The voltage across R_L is computed thus

$$\text{Current } I_L = \frac{V}{R_L + r}$$

$$= \frac{16.8}{5 + 5.045} = \textbf{1.672 mA}$$

$$\text{Voltage across } R_L = I_L \times R_L$$

$$= 1.672 \times 5 = \textbf{8.4 V}$$

Note: Points of interest

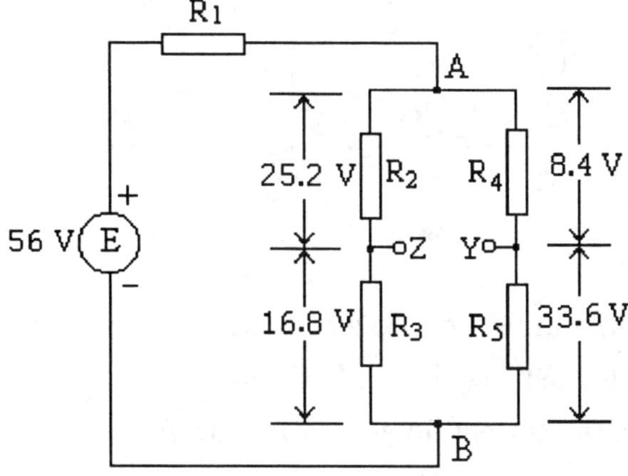

Fig. 2.20-0(d)

From the foregoing calculations we summarize:

<u>Refer to Fig. 2.20-0(d)</u>

$$\text{Voltage across } R_3 = V_{R3} = \underline{16.8 \text{ V}}$$

$$\text{That is, } V_Z = \textbf{\underline{16.8 V}} \quad \textbf{\underline{with respect to B}}$$

$$\text{Voltage across } R_5 = V_{R5} = \underline{33.6 \text{ V}}$$

$$\text{That is, } V_Y = \textbf{\underline{33.6 V}} \quad \textbf{\underline{with respect to B}}$$

V_Y is more positive with respect to V_Z

$$\text{Open-circuit voltage across } \textbf{Y Z} = V_{YZ}$$

$$= 33.6 - 16.8 = \textbf{\underline{16.8 V}}$$

$$\text{With } R_L \text{ connected , } I_L = \textbf{\underline{1.672 mA}}$$

$$\text{and} \quad V_L = \textbf{\underline{8.4 V}}$$

Direction of current is $Y \rightarrow Z$

Example 2.21

A distribution network is arranged as in Fig. 2.21-0. Determine the magnitude and direction of current in the 24 Ω load connected across A D using star to delta transformation. Assume that the internal resistances of the generators are negligible.

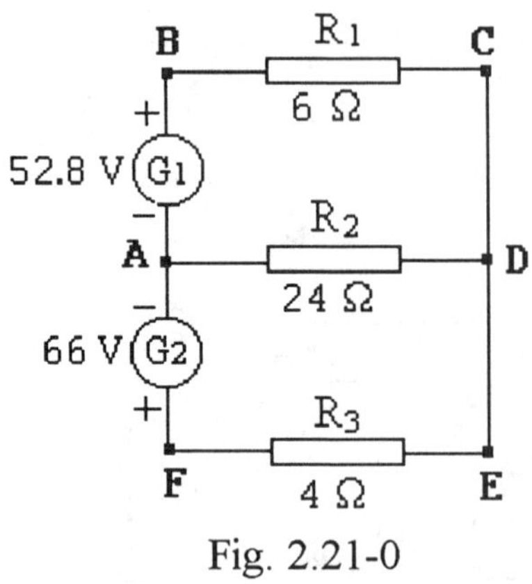

Fig. 2.21-0

Solution

(1) Convert the star network to a delta network as shown in Fig. 2.21-0(a).

(2) Calculate the delta equivalent resistance.

(3) Determine the current flowing from D to A.

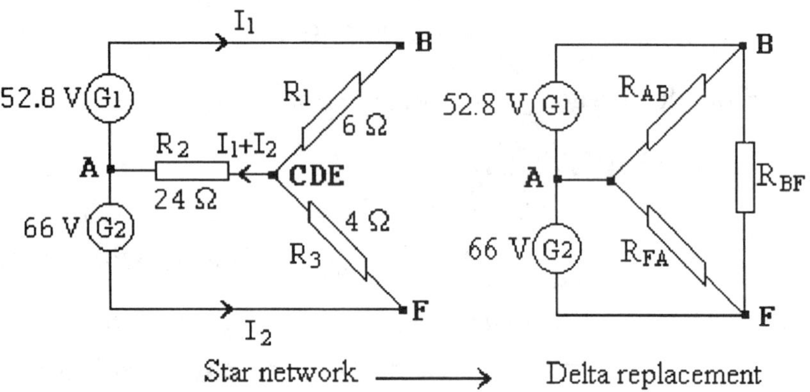

Fig. 2.21-0(a). Star-delta transformation

Refresher – Star to delta transformation

To replace a star network by a delta network divide the three products of all pairs of impedances by the opposite impedance of the star.

$$R_{AB} = \frac{R_2 R_1 + R_1 R_3 + R_3 R_2}{R_3} = \tfrac{1}{4}[(24 \times 6) + (6 \times 4) + (4 \times 24)]$$

$$= \tfrac{1}{4}[(144 + 24 + 96)]$$

$$= \tfrac{1}{4}[264] \qquad = \underline{66\ \Omega}$$

(contd)

$$R_{BF} = \frac{R_1R_3 + R_3R_2 + R_2R_1}{R_2} = \frac{1}{24}[264] = \underline{11\ \Omega}$$

$$R_{FA} = \frac{R_3R_2 + R_2R_1 + R_1R_3}{R_1} = \frac{1}{6}[264] = \underline{44\ \Omega}$$

Current flowing from B to A $= I_1$

$$I_1 = G_1/R_{AB} = {}^{52.8}/66 = \underline{0.8\ A}$$

Current flowing from F to A $= I_2$

$$I_2 = G_2/R_{FA} = {}^{66}/44 = \underline{1.5\ A}$$

Current flowing from D to A $= I_1 + I_2$

$$= 0.8 + 1.5 = \underline{\textbf{2.3 A}}$$

Hence, current flowing in 24 Ω load is 2.3 A from D to A

Example 2.22

A two-wire ring main is represented by the single-line diagram shown in Fig. 2.22-0. The cross-sectional area of each conductor is 0.288 in^2 and the resistivity is 0.8 $\mu\Omega$ - in at the working temperature. Determine the current distribution and the voltage position of minimum p.d.

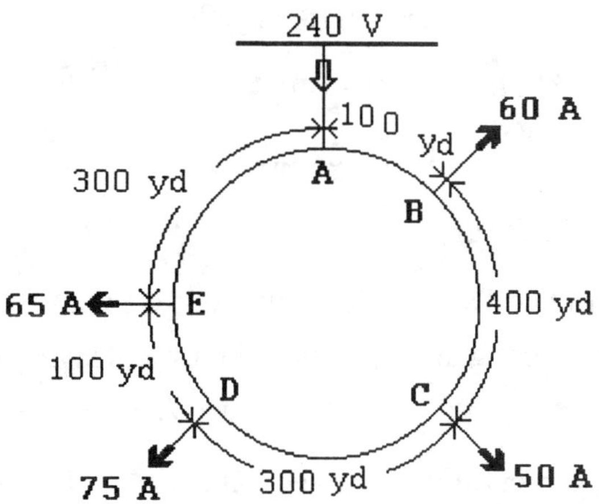

Fig. 2.22-0

Solution

It is convenient to calculate for 100 yd of length of a single conductor since all distances from the source are factors of 100. Thus

$$R = \rho\,{}^{\ell}/a$$

(contd)

$$\text{where } \rho = 0.8 \ \mu\Omega\text{-in}$$

$$\ell = 100 \text{ yd} \times 36 \text{ in}$$

$$a = 0.288 \text{ in}^2$$

$$\therefore \quad R = \frac{0.8 \times 10^{-6} \times 100 \times 36}{0.288} = \underline{0.01 \ \Omega}$$

Resistance of section A B $= 2 \times 0.01 = \underline{0.02 \ \Omega}$

Resistance of section B C $= 2 \times 400/100 \times 0.01 = \underline{0.08 \ \Omega}$

Resistance of section C D $= 2 \times 300/100 \times 0.01 = \underline{0.06 \ \Omega}$

Resistance of section D E $= 2 \times 100/100 \times 0.01 = \underline{0.02 \ \Omega}$

Resistance of section E A $= 2 \times 300/100 \times 0.01 = \underline{0.06 \ \Omega}$

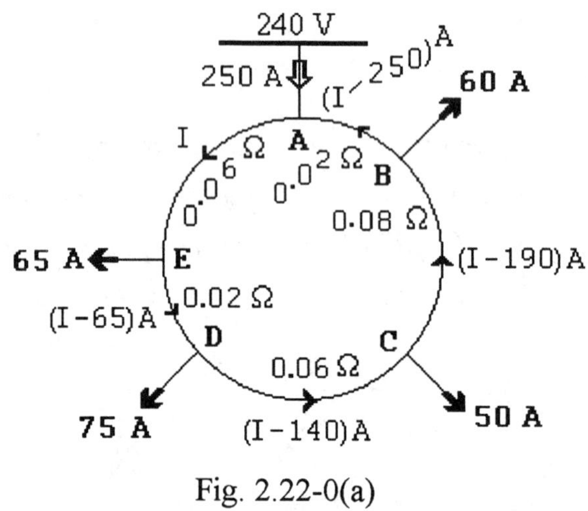

Fig. 2.22-0(a)

<u>Refer to Fig. 2.22-0(a)</u>

Applying Kirchhoff's Laws (current law), we get

Current entering at A $=$ sum of current leaving points B, C, D and E

$$= 60 + 50 + 75 + 65 = \mathbf{250 \ A}$$

Let I be the current flowing in section A to E. Then the current distribution in each section is as shown in Fig. 2.22-0(a).

Applying Kirchhoff's Laws (voltage law)

to closed loop AEDCBA, we get

$$0.06 \ I + 0.02(I - 65) + 0.06(I - 140) + 0.08(I - 190) + 0.02(I - 250) = 0$$

$$0.06 \ I + 0.02 \ I - 1.3 + 0.06 \ I - 8.4 + 0.08 \ I - 15.2 + 0.02 \ I - 5 = 0$$

$$0.24 \ I - 29.9 = 0$$

$$I = 29.9/0.24$$

$$= \mathbf{\underline{124.58 \ A}}$$

(contd)

Substituting for I , we get the current distribution in each section as shown in Fig. 2.22-0(b).

Note the direction of current flow in sections A B, B C, and C D.

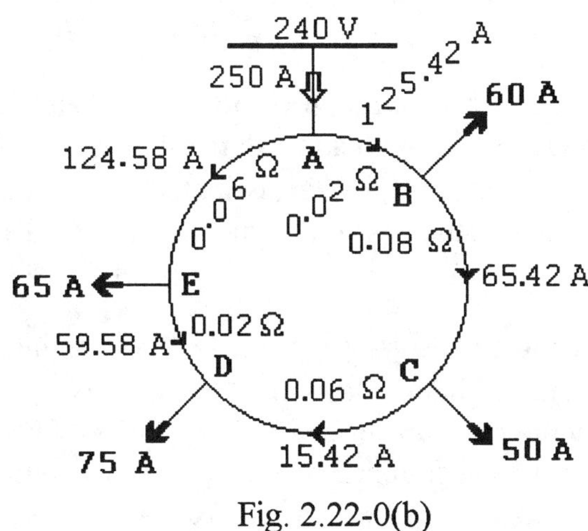

Fig. 2.22-0(b)

<u>Refer to Fig. 2.22-0(b)</u>

It is obvious that the point of minimum p.d is at D.

Voltage drop in sections

A B	=	125.42 x 0.02	= 2.508 V
B C	=	65.42 x 0.08	= 5.234 V
C D	=	15.42 x 0.06	= 0.925 V
Voltage drop from A to D			= 8.667 V
∴ Voltage at point D = 240 – 8.67			= **231.33 V**

Alternatively

Voltage drop in sections

A E	=	124.58 x 0.06	= 7.475
E D	=	59.58 x 0.02	= 1.192
Voltage drop from A to E			= 8.667 V
Voltage at point D = 240 – 8.67			= **231.33 V**

1. Two circuits, A and B, are connected in parallel to a 25 V accumulator, which has an internal resistance of 0.25 Ω. Circuit A consists of two resistors, 10 Ω and 5 Ω, connected in series. Circuit B consists of two resistors, 6 Ω and 4 Ω, connected in series. Determine the current flowing in and the p.d across each of the four resistors.
    [Ans. A-1.6 A, 16 V, 8 V; B- 2.4 A, 14.4 V, 9.6 V]

2. A circuit consists of three resistors, 36 Ω, 18 Ω and 12 Ω, joined in parallel and connected in series with a fourth resistance. The circuit is supplied at 60 V and it is found that 36 W is dissipated in the 12 Ω resistor. Determine the value of the fourth resistor and the total power dissipated in the group.
    [Ans. 11.33 Ω, 208 W]

3. A battery having a constant e.m.f of 245 V and an internal resistance of 0.1 Ω is connected in parallel with a dynamo, the p.d of which falls uniformly on open circuit, from 250 V to 230 V, when delivering 100 A. Calculate the current supplied by the dynamo when the total current taken by an external load is (a) zero and (b) 100 A.
    [Ans. (a) 16.67 A, (b) 50 A]

4. Two batteries, A and B, are connected in parallel and a 5 Ω resistor is connected across the battery terminals. The e.m.f and internal resistance of battery A are 4 V and 2 Ω, respectively, and the corresponding values for battery B are 1.0 V and 3 Ω, respectively. Calculate the current flowing in each battery and in the 5 Ω resistor.
    [Ans. A- 0.871 A, B- 0.419 A; 0.452 A]

5. A Wheatstone bridge is supplied from a 10 V battery of negligible internal resistance. A galvanometer of 10 Ω resistance is connected to the center arm BD of the bridge. The resistances of the arms AB, BC, CD and DA are 5 Ω, 3 Ω, 2 Ω and 5 Ω, respectively. Find the current in the center arm by (a) Kirchhoff's Laws and (b) Thevenin's Theorem.
    [Ans. 0.0671 A]

6. Two batteries E_1 and E_2 having e.m.fs of 200 V and 190 V have internal resistances of 0.2 Ω and 0.5 Ω, respectively. To charge the batteries the circuit shown in Fig. 2.P-6 was employed.

(contd)

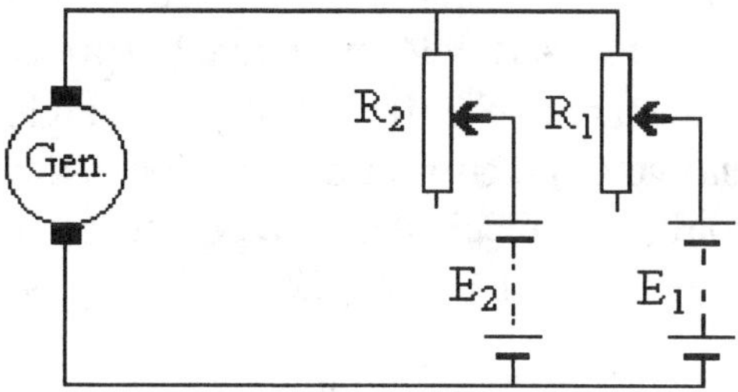

Fig. 2.P-6

The values or R_1 and R_2 are 9.6 Ω and 14.3 Ω, respectively, and the
generator has an e.m.f of 220 V and an internal resistance of 0.1 Ω.
Calculate the values of the charging currents and the current supplied by
the generator. *[Ans. 2 A, 2 A; 4 A]*

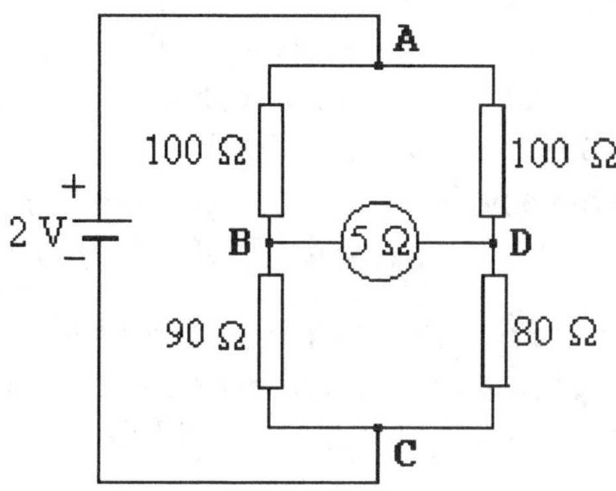

Fig. 2.P-7

7. Determine the direction and magnitude of the current in the galvanometer
arm of the Wheatstone bridge circuit shown in Fig. 2.P-7 using
(a) Thevenin's Theorem and (b) delta-star transformation.
[Ans. 0.608 mA from B to D]

8. Two separately excited d.c. generators have armature resistances of 0.1 Ω
and 0.2 Ω and generate e.m.fs of 24 V and 25 V, respectively. The generators are connected in parallel such that their positive terminals are linked
by a bus bar of negligible resistance and each negative terminal is 'earthed'
(grounded) through a 0.3 Ω resistor. Calculate the current which would
circulate through the machines and determine their terminal voltages.
If a load of 6 Ω is connected between the bus bar and earth (ground),
calculate the current which will flow through the load and the terminal
voltage of each generator.
[Ans. 1.11 A, 24.4 V; 3.93 A, 23.58 V]

9. A two-wire distributor is fed at 230 V from end A and loads of 40 A, 60 A, and 30 A are connected at points B, C and D. Loads B, C and D are 100, 250 and 470 yd, respectively from point A. The distributor has a resistance of 0.01 Ω/100 yd of single conductor. Sketch a diagram showing the current distribution and determine the voltage at each load point; hence calculate the power loss in the distributor.

[Ans. B- 227.4 V; C- 224.7 V; D- 223.4 V; 620.6 W]

10. A two-wire distributor 400 yd long is loaded with 100 A, 108 A, 50A and 25 A, respectively. The load points are 100 yd, 250 yd, 300 yd and 400 yd from the feeding point. The resistivity of the conductors is 0.6 $\mu\Omega$-in at normal working temperature. Calculate the cross-sectional area of each conductor.

[Ans. 0.2 in²]

11. A ring main distributor 500 yd long is fed at A and loads of 100 A and 150 A are connected to points B and C, respectively. Load B is 200 yd from A, and C is 250 yd from B. Calculate (a) the current distribution in the ring main and (b) the supply voltage and p.d across load B if the p.d across C is to be maintained at 200V.
(Resistance of cable is 0.02 Ω/100 yd/single conductor.)

[Ans. (a) 75 A, 25 A, 175 A. (b) 203.5 V, 197.5V]

12. The diagram in Fig. 2.P-12 shows a three-wire d.c system on load, with resistance of each section given in ohms. Calculate the section currents and voltages across the loads DD and EE.

[Ans. 236.1 V, 209.6 V]

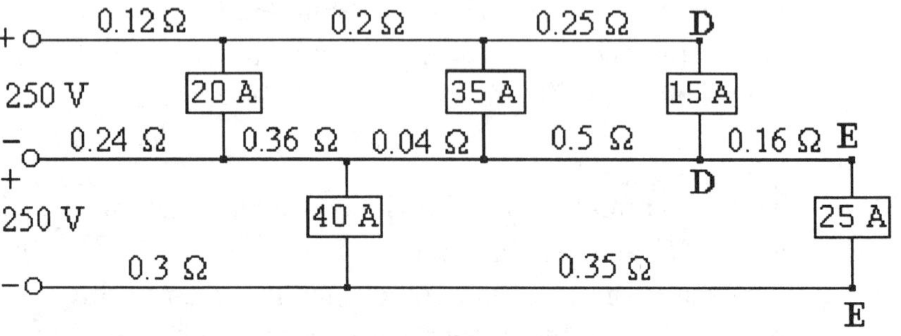

Fig. 2P-12

13. A two-wire distributor 400 yd long is fed from both ends at 200 V and is loaded with 120 A, 150 A and 100 A at distances of 100 yd, 250 yd and 325 yd, respectively, from the left-hand side (LHS). The resistance of a single conductor is 0.03 Ω/100 yd. Calculate (a) the current distribution in the distributor and (b) the voltage across each load.

[*Ans. (a) From LHS 165 A, 45 A, from RHS 205 A, 105 A*;
(b) 190.1 V, 186.05 V, 190.77 V]

14. A ring main has a pair of uniform conductors 600 yd long and is fed at 240 V. Loads of 60 A, 40 A and 50 A are connected at distances of 150 yd, 350 yd, and 440 yd, respectively, in one direction from the feeding point. Calculate (i) the current in each section of the ring main, (ii) the voltage at each load, and (iii) the power loss in the conductors. (Resistance of conductor is 0.05 Ω/1000 yd.)

[*Ans. (i) 75 A,15 A, 25 A, (ii) 238.9 V, 238.6 V, 238.8V,*
(iii) 184.5 W]

CHAPTER 3

MAGNETISM AND ELECTROMAGNETISM

Introduction

A magnetic circuit consists essentially of an interconnected set of branches of ferrous magnetic materials in which a magnetic flux is established. The magnetic flux is established either by permanent magnetism in the material or by an electric current flowing in a coil which wraps around part of the material. The flux is in fact stationary, but, for convenience, it is assumed to flow around the magnetic circuit. With this assumption, the magnetic circuit is analogous with the electric circuit.

Consider the two circuits shown in Fig. 3.0-0.

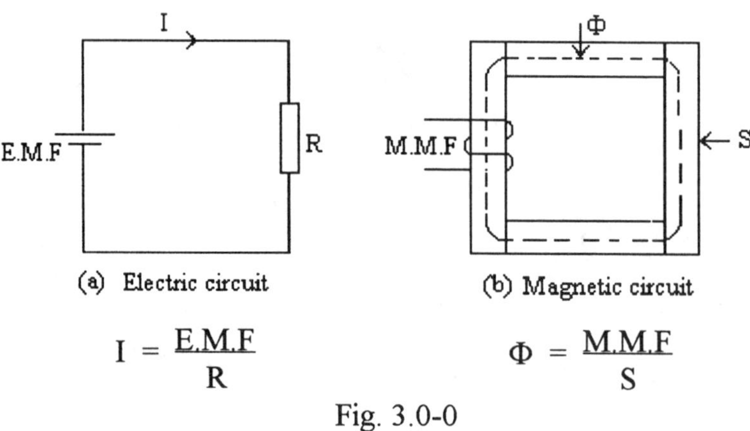

$$I = \frac{E.M.F}{R} \qquad \Phi = \frac{M.M.F}{S}$$

Fig. 3.0-0

Fig. 3.0-0(a). The E.M.F (volts) produces a current I (amperes) flowing through the resistance R (ohms) in the electric circuit.

Fig. 3.0-0(b). The M.M.F [ampere turns (AT)] produces a flux Φ (webers) through the reluctance S (AT/Wb) in the magnetic circuit.

E.M.F (V) is analogous with M.M.F (AT).

I (A) is analogous with Φ (Wb).

R (ohm) is analogous with S (AT/Wb).

Nomenclature, Formulae and Symbols

1. *Magnetizing force* (M.M.F) is the number of ampere-turns applied to a magnetic circuit.

$$M.M.F = \text{amperes} \times \text{turns} = I\,N \quad [\text{unit, AT or A}]$$

2. *Magnetizing force* (H) is the magnetomotive force (m.m.f) gradient across a magnetic circuit [unit, AT/m or A/m].

$$H = \frac{\text{magnetomotive force}}{\text{length of magnetic circuit}} = \frac{m.m.f}{\ell}\,(\text{AT/m})$$

3. *Magnetic flux* (Φ) may be described as the total amount of magnetic lines of force or magnetic field in a magnetic circuit.

$$\Phi = \text{flux density} \times \text{area} = B \times A\,(\text{Wb})$$

4. *Magnetic flux density* (B) is the magnetic flux within the cross-sectional area of the magnetic material.

$$B = \frac{\text{magnetic flux}}{\text{cross-sectional area}} = \frac{\Phi}{A}\ \text{tesla}\,(\text{Wb/m}^2)$$

5. *Permeability of free space* (μ_o) is given as the ratio B/H for a non-magnetic medium or in a vacuum or free space. It has a numerical value of $4\pi \times 10^{-7}$ henry/meter

$$\mu_o = {}^{B}\!/\!_{H} \text{ in a vacuum} = 4\pi \times 10^{-7}\ \text{H/m}$$

6. *Relative permeability* (μ_r) is defined as the ratio of flux density produced in the material by a magnetizing force H to that produced in a non-magnetic medium by the same magnetizing force H.

$$(\mu_r = 1 \text{ for free space})$$

7. *Reluctance* (S) is that property of a material which opposes the increase of magnetic flux.

$$S = \frac{\ell}{\mu_o\,\mu_r\,A}\quad \text{AT/Wb}$$

$$\text{or}\quad S = \frac{M.M.F}{\Phi}\quad \text{AT/Wb}$$

8. *Permeance* (Λ) is the reciprocal of reluctance, which is analogous to conductance in an electric circuit.

$$\Lambda = {}^{1}\!/\!_{S}\quad [\text{analogous to conductance, } G = {}^{1}\!/\!_{R}]$$

[The symbol Λ is the Greek letter lambda in uppercase.]

9.   *Self-inductance* (L) of a coil is given by

$$L = \frac{\mu_o \mu_r A N^2}{\ell} \quad \text{(henry)}$$

where μ_o = permeability of free space = $4\pi \times 10^{-7}$

μ_r = relative permeability of the material

A = cross-sectional area of core in meters^2

ℓ = length of coil in meters

also

$$L = N \, d\Phi/di \quad \text{(henry)}$$

10.   *Mutual inductance* (M) is that property resulting when two coils (primary and secondary) are coupled such that a change in the magnetic flux in one coil causes an e.m.f to be induced in the other.

$$M = \frac{\text{change of flux linkages in secondary}}{\text{change of current in primary}} \quad \text{(H)}$$

11.   *Induced e.m.f (e) or e.m.f (e) of self-inductance* is given as follows:

In a coil, $e = -N \, d\Phi/dt$ volts

where   N = number of turns affected by the rate of change of flux , $d\Phi/dt$ (Wb/sec)

The minus sign indicates that the induced e.m.f is in opposite polarity to the source creating it.

In a circuit, $e = -L \, di/dt$

$$-L \, di/dt = -N \, d\Phi/dt$$

$$L = N \, d\Phi/di \quad \text{[refer to (9)]}$$

12.   *Force* (F) is the force exerted on a current-carrying conductor which lies at right angles to the magnetic field and is given by

$$F = B I \ell \quad \text{newtons}$$

where B = flux density in teslas (Wb/m^2)

I = current in amperes

ℓ = length of straight conductor in meters

When the conductor lies at an angle θ to the direction of the magnetic field, the expression is modified to

$$B = B I \ell \sin \theta \quad \text{newtons}$$

13. *Induced e.m.f (e) in a stationary field and a conductor in motion* is given as follows:

If a conductor of length ℓ lies in a uniform magnetic field and is moved through a distance d meters in time t seconds at right angles to the magnetic field, then the e.m.f induced in the conductor is given by

$$e = B\,\ell\,v \text{ volts} \quad [v = \text{velocity} = \text{distance}/\text{time} = d/t]$$

Should the direction of motion of the conductor be at an angle θ to the magnetic field, then the equation becomes

$$e = B\,\ell\,v \sin\theta \text{ volts}$$

14. *Energy stored in a magnetic field* is given by

$$B^2/2\mu_0 \quad (J/m^3)$$

15. *Energy stored in an inductor* is given by

$$\tfrac{1}{2} L I^2 \quad \text{(joules)}$$

16. *Pull (P) or attractive force* between magnetized surfaces

$$= \frac{B^2 A}{2\mu_0} \quad \text{(newtons)}$$

Area (A) of attracting faces is assumed equal.

17. *Leakage coefficient or magnetic fringing* is defined as

$$\frac{\text{total flux}}{\text{useful flux}}$$

18. *Coefficient of coupling (k)* is given by

$$k = M/\sqrt{(L_1 L_2)}$$

19. *Time constant* (T) is given by

$$T = L/R \text{ seconds}$$

where L is in henrys

and   R is in ohms

20. *Hysteresis loss* is given by

$$P_h \propto f (B_{max})^n \quad (n \text{ lies between } 1.6 \text{ and } 2.2)$$

21. *Inductance of two coils in series, aiding (cumulatively coupled)*, is

$$L = L_1 + L_2 + 2M$$

22. *Inductance of two coils in series, opposing (differentially coupled)*, is

$$L = L_1 + L_2 - 2M$$

Note: The unit of magnetic flux (Φ) is the weber.

The unit of magnetic flux density (B) is the tesla.

$$\text{Magnetic flux } \Phi = \text{flux density (B)} \times \text{area (A)}$$

$$\Phi \text{ (weber)} = B \text{ (tesla)} \times A \ (m^2)$$

$$\therefore \quad B \text{ (tesla)} = \frac{\Phi \text{ (weber)}}{A \ (m^2)}$$

$$\textbf{1 tesla} = \textbf{1 weber/m}^2$$

23.   *Faraday's Law*   states that

"an e.m.f is induced in a circuit whenever the flux linkages with that circuit are changed."

24.   *Neumann's Law*   states that

"the magnitude of the e.m.f induced in a circuit is proportional to the rate of change of magnetic flux linkages with that circuit."

25.   *Lenz's Law*   states that

"the induced e.m.f tends to set up a current in a direction that opposes the change in flux which produces the induced e.m.f"

or   "the direction of the induced e.m.f is always such as to oppose the source which produces it."

Example 3.1

The flux linking an air-cored coil of 800 turns changes from 40 μWb to 80 μWb in 0.002 sec. Calculate the value of the induced e.m.f in the coil.

Solution

$$\text{Induced e.m.f}, e = N \times \frac{\text{change of flux linkages}}{\text{change of time}}$$

$$e = N \, d\Phi/dt$$

$$= 800 \times \frac{(80 - 40) \times 10^{-6}}{0.002}$$

$$= 800 \times \frac{40 \times 10^{-6}}{2 \times 10^{-3}}$$

$$= 800 \times 20 \times 10^{-3} = \underline{\textbf{16 V}}$$

[Note: Polarity has been omitted and will be introduced in later examples.]

Example 3.2

*A current-carrying conductor lies at right angles to a uniform magnetic
field of flux density 0.3 T. Calculate the force in newtons per meter of
length of the conductor when the current in it is 200 A.*

Solution

$$\text{Force, F} = B\,I\,\ell \text{ newtons}$$

where B = flux density in teslas

I = current in amperes

ℓ = length of conductor in meters

$$\therefore \qquad F = 0.3 \times 200 \times 1 \qquad = \underline{\textbf{60 newtons}}$$

Example 3.3

*A conductor carrying a current lies at right angles to a uniform magnetic
field of flux density 0.5 T. If the force acting on the conductor is 400 N
per meter of length, calculate the current in the conductor.*

Solution

$$\text{Force, F} = B\,I\,\ell \text{ newtons}$$

$$\text{Current, I} = {}^{F}/B\,\ell \text{ amperes}$$

$$= \frac{400}{0.5 \times 1} \qquad = \underline{\textbf{800 A}}$$

Example 3.4

*A conductor of length 0.5 m moves perpendicular to a magnetic field at a
velocity of 30 m/s. Calculate the average value of the induced e.m.f in the
conductor if the magnetic flux density is 0.4 T.*

Solution

$$\text{Induced e.m.f, } e = B\,\ell\,v \sin\theta \text{ volts}$$

Since the conductor is perpendicular to the magnetic flux,

$$\theta = 90^{\circ}$$

$$\therefore \qquad e = 0.4 \times 0.5 \times 30 \ \sin 90^{\circ}$$

$$= 0.4 \times 0.5 \times 30 \times 1 \qquad = \underline{\textbf{6.0 V}}$$

Example 3.5

The average value of an e.m.f induced in a conductor of length 0.3 m is 0.75 V. The conductor moves at an angle θ^o with reference to the magnetic field and at a velocity of 20 m/s. If the magnetic flux density is 0.5 T, calculate the direction of movement of the conductor.

Solution

$$\text{Induced e.m.f, } e = B\,\ell\,v\,\sin\theta \text{ volts}$$
$$\sin\theta = {}^e/B\,\ell\,v$$
$$= {}^{0.75}/(0.5 \times 0.3 \times 20) = \underline{\mathbf{0.25}}$$
$$\theta = \sin^{-1} 0.25 = \underline{\mathbf{14.48^o}} \text{ or } \underline{\mathbf{14^o\ 29'}}$$

Example 3.6

A coil of 400 turns is wound uniformly over a non-magnetic ring of mean circumference 0.3 m and cross-sectional area 600 mm². If the current through the coil is 6 A, calculate (a) the magnetic field strength, (b) the flux density and (c) the total flux.

Solution

(a) Magnetizing field strength , $H = {}^{m.m.f}/\ell = {}^{I\,N}/\ell$

Mean circumference, $\ell = 0.3$ m

$\therefore \qquad H = \dfrac{6 \times 400}{0.3} = \underline{\mathbf{8000\ AT/m}}$

(b) Flux density, $B = \mu_o H \qquad [{}^B/H = \mu_o]$

$= 4\,\pi \times 10^{-7} \times 8000 = \underline{\mathbf{0.01\ T}}$

(c) Total flux , Φ = flux density (B) x area (A)

$= 0.01 \times 600 \times 10^{-6} = \underline{\mathbf{6\,\mu Wb}}$

Example 3.7

Determine the magnetomotive force (m.m.f) required to establish a flux of 0.025 Wb across the air gap shown in Fig. 3.7-0. The effective cross-sectional area of the material is 200 cm².

Solution

We are required to find the m.m.f from the data given.

Flux, $\Phi = 0.025$ Wb

Length of air gap = 3.5 mm   = 0.0035 m

(contd)

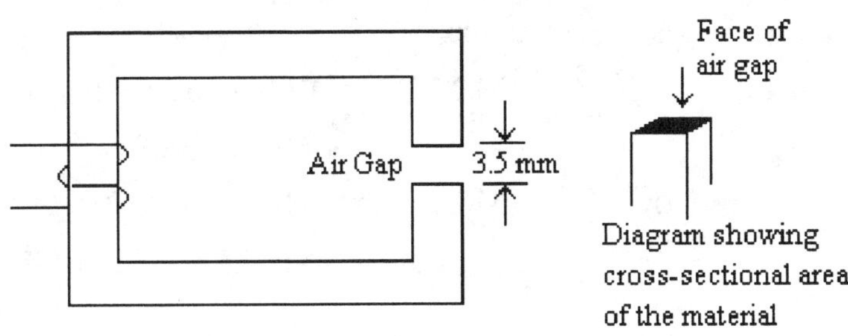

Fig. 3.7-0

Cross-sectional area of air gap = cross-sectional area of material

$$= 200 \text{ cm}^2 \quad = 200 \times 10^{-4} \quad = 0.02 \text{ m}^2$$

Permeability of free space, $\mu_o = 4\pi \times 10^{-7}$

First calculate the flux density B.

Second calculate the magnetic field strength H for the air gap.

Finally, required m.m.f = H (AT/m) x length of air gap (m).

$$B \quad = \Phi/A \ = \ ^{0.025}/0.02 \ = \underline{1.25 \text{ T}}$$

$$H \quad = \ ^B/\mu_o \ = \ ^{1.25}/(4\pi \times 10^{-7})$$

$$= \underline{995223 \text{ AT/m}}$$

$$\therefore \text{ Required m.m.f} \quad = H \times \ell = 995223 \text{ (AT/m)} \times 0.0035 \text{ m}$$

$$= \underline{\textbf{3483 AT}}$$

Example 3.8

A ring fabricated from cast steel has a cross-sectional area of 400 mm²
and a mean circumference of 500 mm. The ring is uniformly wound with
200 turns of insulated wire. Calculate

    (a) the reluctance of the ring

    (b) the current required to set up a flux of 600 μWb in the ring

        [Take μ_r = 400]

Solution

(a)

$$\text{Reluctance, S} \quad = \ ^\ell/\mu_o\mu_r A \ \text{ AT/Wb} \quad [\text{also } S = \ ^{M.M.F}/\Phi]$$

$$= \frac{500 \times 10^{-3}}{4\pi \times 10^{-7} \times 400 \times 400 \times 10^{-6}}$$

$$= \frac{500 \times 10^{6}}{4\pi \times 4 \times 4} \qquad = \underline{\textbf{2.488 x 10}^6 \textbf{ AT/Wb}}$$

(contd)

(b)   Also   reluctance, $S = \text{M.M.F}/\Phi$

$\therefore$ $\text{M.M.F} = \Phi S = 600 \times 10^{-6} \times 2.488 \times 10^{6}$

$= 600 \times 2.488 \qquad = \underline{1492.8 \text{ AT}}$

Now $\text{M.M.F} = I\,N = 1492.8 \text{ AT}$

Current required, $I = \text{M.M.F}/N = 1492.8/200 = \underline{\textbf{7.464 A}}$

Alternatively

$$H = B/\mu_o\mu_r$$

$$B = \Phi/A \qquad = \frac{600 \times 10^{-6}}{400 \times 10^{-6}} = \underline{1.5 \text{ T}}$$

$$H = 1.5/(4\pi \times 10^{-7} \times 400) = \underline{2985.67 \text{ AT/m}}$$

$$\text{M.M.F} = H \times \ell = 2985.67 \times 0.5 = \underline{1492.8 \text{ AT}}$$

$$I = \text{M.M.F}/N = 1492.8/200 = \underline{\textbf{7.464 A}}$$

Example 3.9

A mild steel ring has a cross-sectional area of 20 cm² and a mean diameter of 50 cm. The ring is uniformly wound with 500 turns of wire and a current of 2 A produced a flux of 2.5 mWb. Calculate the relative permeability of the material.

Solution

$$\text{Flux, } \Phi = \text{M.M.F}/S$$

$$\text{Reluctance, } S = \ell/\mu_o\mu_r A$$

$$\therefore \quad \Phi = \text{M.M.F} \times \mu_o\mu_r A/\ell$$

$$\therefore \text{ Relative permeability, } \mu_r = \frac{\ell\,\Phi}{\text{M.M.F} \times \mu_o A}$$

Mean length, $\ell = \pi\,d = \pi \times 50 \times 10^{-2} \text{ m}$

Flux, $\Phi = 2.5 \times 10^{-3} \text{ Wb}$

$\text{M.M.F} = I \times N = 2 \times 500 \text{ AT}$

$\mu_o = 4\pi \times 10^{-7}$

Area, $A = 20 \times 10^{-4}$

$$\therefore \quad \mu_r = \frac{\pi \times 50 \times 10^{-2} \times 2.5 \times 10^{-3}}{2 \times 500 \times 4\pi \times 10^{-7} \times 20 \times 10^{-4}} = \frac{50 \times 2.5 \times 10^{-5}}{1000 \times 80 \times 10^{-11}}$$

$$= \frac{50 \times 2.5 \times 10^{6}}{8 \times 10^{4}} = \frac{125 \times 10^{2}}{8} = \underline{\textbf{1562.5}}$$

Example 3.10

The air gap in a magnetic circuit is 0.2 cm long and has a cross-sectional area of 30 cm². Calculate the flux density in the air gap if the magnetomotive force (m.m.f) is 800 AT.

The magnetic circuit with flux path Φ and associated air gap is shown in Fig. 3.10-0.

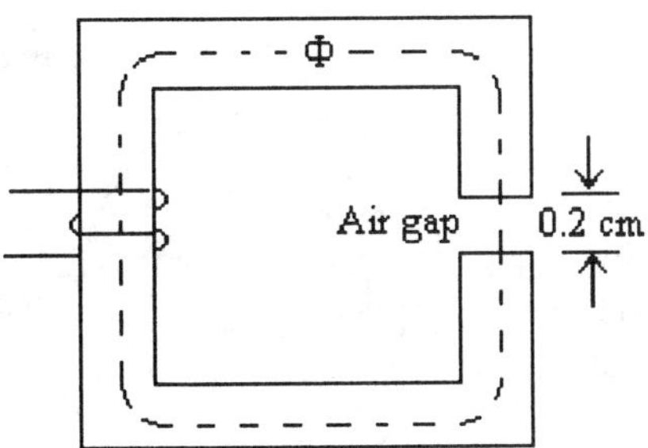

Fig. 3.10-0

Solution

$$\frac{B}{H} = \mu_o\mu_r$$

$$\therefore \quad \text{Flux density,} \quad B = \mu_o\mu_r\,H$$

where for the air gap, $\mu_o = 4\pi \times 10^{-7}$

and $\qquad \mu_r = 1$

$$\text{Magnetizing force,} \quad H = \frac{IN}{\ell}$$

$$\ell = \text{length of air gap}$$

$$= 0.2 \times 10^{-2}\ \text{m}$$

$$\therefore \quad B = \frac{4\pi \times 10^{-7} \times 1 \times 800}{0.2 \times 10^{-2}}\ \text{Wb/m}^2$$

$$= \frac{4\pi \times 80 \times 10^{-6}}{2 \times 10^{-3}}$$

$$= 2\pi \times 80 \times 10^{-3} \qquad = \underline{\textbf{502.4 mWb/m}^2}$$

Example 3.11

A cast steel ring of mean circumference 0.1π m and cross-sectional area 4 cm² is uniformly wound with 1500 turns of wire. Find the current required to produce a flux of 480 μWb. in the ring. Refer to Fig. 3.11-0.

[Use Fig. 3.11-1 B/H curves.]

Solution

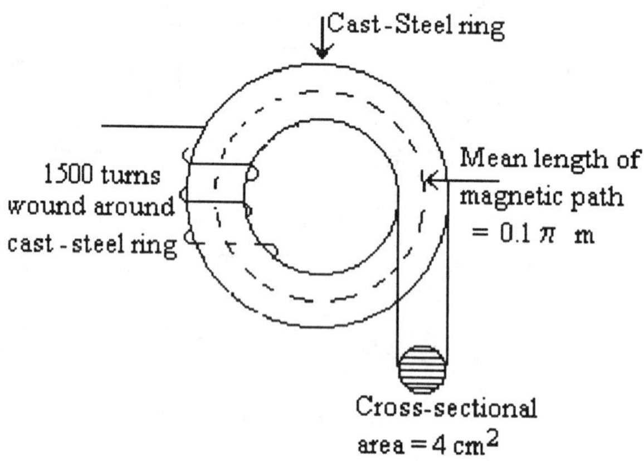

Fig. 3.11-0

Flux density, $B = \Phi/A$

$$= \frac{480 \times 10^{-6}}{4 \times 10^{-4}} \qquad = \underline{1.2\ T}$$

From the B/H curves in Fig. 3.11-1 the corresponding magnetizing force H for this flux density (1.2 T) is

$$H = 1550\ AT/m \ \text{(approximately)}$$
$$M.M.F = H \times \ell$$
$$= 1550 \times 0.1\pi\ (\pi = 3.14) \qquad = \underline{486.7\ AT}$$

$\therefore \qquad$ Current, $I = M.M.F/N$

$$= 486.7/1500 \qquad\qquad = \underline{\mathbf{0.324\ A}}$$

Variation of Flux Density with Magnetizing Field Strength

for Stalloy, Cast Steel and Cast Iron

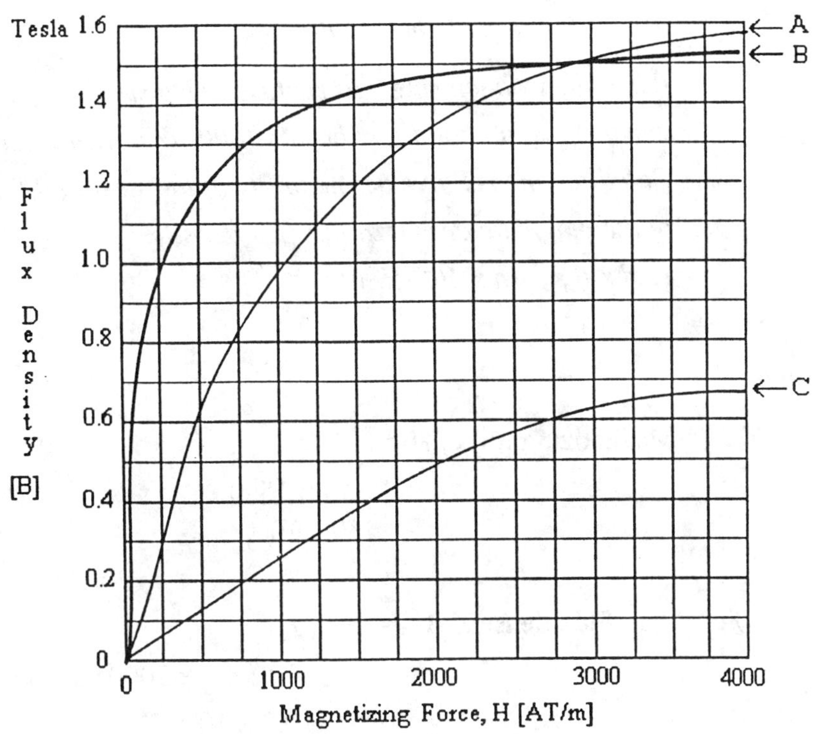

Fig. 3.11-1. B/H curves

A: Cast Steel

B: Stalloy

C: Cast Iron

109

Example 3.12

    (a)   *A coil of 1500 turns is uniformly wound on a circular wooden ring which has a mean circumference of 30 cm and a cross-sectional area of 4 cm². Calculate*

        (i)   *the flux density in the ring when the coil carries a current of 0.4 A*

        (ii)   *the flux in the ring in webers*

    (b)   *When the wooden ring was replaced by a steel ring of the same dimensions, the total flux became 600 μWb with a current of 0.4 A. Calculate the relative permeability of the steel and the reluctance of the magnetic circuit at this flux density.*
[Take $\mu_o = 4\pi \times 10^{-7}$ H/m]

Solution

(a)   *(i)*   Magnetizing force, $H = I\,N/\ell$

$$= \frac{0.4 \times 1500}{30 \times 10^{-2}} \qquad = \frac{600}{0.3}$$

$$= \underline{2000\ \text{AT/m}}$$

    (ii)   Flux density, $B = \mu_o \mu_r H$

$$= 4\pi \times 10^{-7} \times 1 \times 2000 \quad [\text{for wood } \mu_r = 1]$$

$$= 4\pi \times 0.2 \times 10^{-3}$$

$$= 2.512 \times 10^{-3} \qquad = \underline{\mathbf{0.002512\ T}}$$

Flux, $\Phi = B \times A$

$$= 2.512 \times 10^{-3} \times 4 \times 10^{-4}\ \text{Wb}$$

$$= 2.512 \times 4 \times 10^{-7} \times 10^{6}\ \mu\text{Wb}$$

$$= 2.512 \times 4 \times 10^{-1} \qquad = \underline{\mathbf{1.0048\ \mu Wb}}$$

(b)   *Steel ring replaces wooden ring*

Magnetizing force, $H = 2000$ AT/m

Flux, $\Phi = 600\ \mu$Wb

Flux density, $B = \Phi/A$

$$= \frac{600 \times 10^{-6}}{4 \times 10^{-4}} \qquad = \underline{1.5\ T}$$

(contd)

$$\text{Now } B/H = \mu_0\mu_r$$

$$\therefore \text{ Relative permeability, } \mu_r = B/\mu_0 H$$

$$= \frac{1.5}{4\pi \times 10^{-7} \times 2000}$$

$$= \frac{1.5 \times 10^3}{2.512} = \mathbf{597}$$

$$\text{Reluctance, } S = \text{M.M.F}/\Phi$$

$$= \text{IN}/\Phi$$

$$= \frac{0.4 \times 1500}{600 \times 10^{-6}} = \mathbf{1.0 \times 10^6 \ AT/Wb}$$

Example 3.13

A section of the armature of a d.c motor has 150 conductors, each of length 95 cm. The conductors are positioned perpendicular to a magnetic field having a flux density of 0.8 T and carry a current of 25 A. If the conductors are situated at a radial distance of 20 cm from the center-line of the armature, calculate

(a)   the force on each conductor

(b)   the torque developed by each conductor

(c)   the torque developed by the section of the armature

Solution

(a)   Force on each conductor is given by the expression

$$\text{Force, } F = B I \ell \text{ newtons [N]}$$

where flux density, $B = 0.8$ T

current, $I = 25$ A

length, $\ell = 0.95$ m

$$\therefore \quad F = 0.8 \times 25 \times 0.95 = \mathbf{19 \ N}$$

(b)   Torque T developed by each conductor is

$$T = \text{force} \times \text{radial distance}$$

$$= 19 \times 20 \times 10^{-2} = \mathbf{3.8 \ N \ m}$$

(c)   Torque T developed by the 150 conductors is

$$T = \text{torque on 1 conductor} \times \text{number of conductors}$$

$$= 3.8 \text{ (N m)} \times 150 = \mathbf{570 \ N \ m}$$

Example 3.14

*An iron ring is wound with a coil of 200 turns and carries a current of
2.0 A. The ring has a cross-sectional area of 3 cm² and a mean diameter
of 25 cm. An air gap of 0.4 mm has been made by a cut across
the section of the ring. If the total flux is 210 μWb, calculate the relative
permeability of the iron, assuming no magnetic leakage.
Refer to Fig. 3.14-0.*

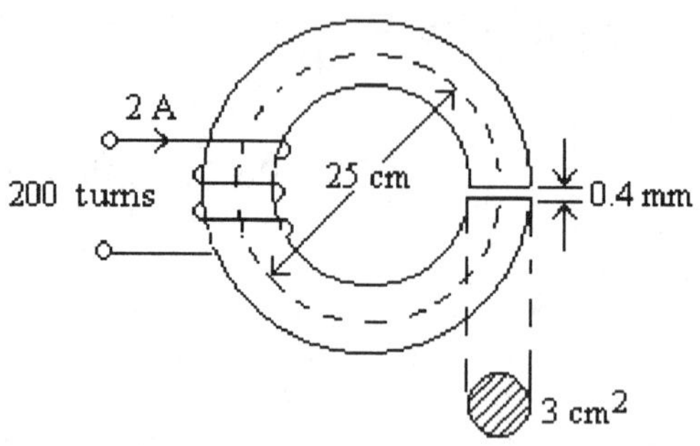

Fig. 3.14-0

Solution

Flux density (B) in Iron and air gap is

$$B = \frac{\text{total flux } (\Phi)}{\text{cross-sectional area of ring (A)}}$$

$$= \frac{210 \times 10^{-6}}{3 \times 10^{-4}}$$

$$= 70 \times 10^{-2} \text{ T} \qquad = \underline{0.7 \text{ T}}$$

For Air gap only

Flux density, $B = \mu_o \mu_r H$

Magnetizing force, $H = {}^{B}/_{\mu_o \mu_r}$ $(\mu_r = 1 \text{ for air gap})$

$$= {}^{0.7}/(4\pi \times 10^{-7}) \quad = \underline{557324.84 \text{ AT/m}}$$

Ampere-turns required for air gap

$$= H \times \ell = 557324.84 \ \times 0.4 \times 10^{-3}$$

$$= \underline{222.93 \text{ AT}}$$

For Iron only

Total M.M.F (AT) produced $= I \times N$ (AT)

$$= 2 \times 200 \qquad = \underline{400 \text{ AT}}$$

M.M.F (AT) available for Iron is the difference between the total M.M.F
(AT) produced and that which is required for the air gap. That is,

M.M.F available for Iron $= 400 - 222.93 \qquad = \underline{177.07 \text{ AT}}$

(contd)

$$\text{Magnetizing force, } H = \frac{\text{M.M.F}}{\ell}$$

$$\text{length, } \ell = \pi \times \text{mean diameter (d)}$$

$$\therefore \quad H = \frac{177.07}{\pi \times 25 \times 10^{-2}} = \underline{225.57 \text{ AT/m}}$$

$$\text{Flux density, } B = \mu_o \mu_r H$$

$$\text{Relative permeability, } \mu_r = \frac{0.7}{4\pi \times 10^{-7} \times 225.57}$$

$$= \frac{0.7 \times 10^7}{2833.16} = \underline{\mathbf{2471}}$$

Example 3.15

(a) *Define the MKS unit of current and show that the permeability of free space $\mu_o = 4\pi \times 10^{-7}$ H/m.*

(b) *A semicircular lifting magnet having a cross-sectional area of 5 cm² supports a load hanging from a straight armature. The length of the magnetic circuit is 20 cm and each air gap is 0.1 mm long. Calculate the m.m.f required to set up a flux of 0.5 mWb in the armature. Three points on the B/H curve are:*

H (AT/m)	400	500	800
B (Tesla)	0.8	1.0	1.2

Evaluate the tractive force the magnet would exert with this magnetization. Assume the magnetic pull to be $B^2/2\mu_o$ N/m². What load, in pounds, could the armature then support?
[1 N = 0.225 lb]

Solution

(a) When two conductors carrying current are situated so that they are parallel to each other, a force is developed between them. If the currents in the conductors are in the same direction, then the force is attractive; if, however, the currents are in opposite direction the conductors will repel each other.

When a force of 2×10^{-7} N/m is maintained on two infinitely long parallel conductors separated by 1 m between centers and situated in a vacuum, the current in the conductors is 1 A. Refer to Fig. 3.15-0.

113

(contd)

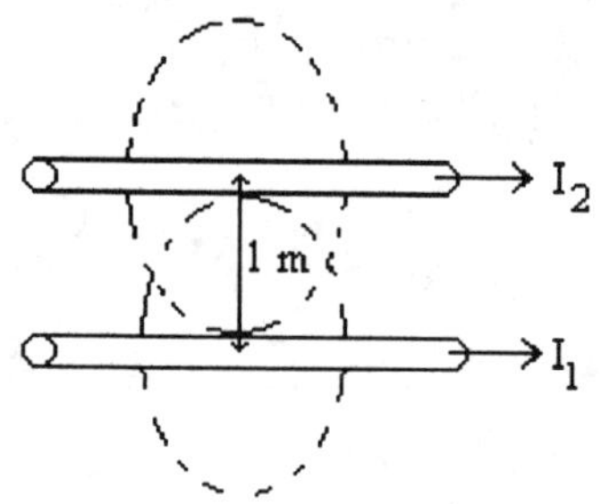

Fig. 3.15-0

Consider a straight section of the infinitely long conductor shown in
Fig. 3.15-1.

$$M.M.F = I \times N = 1 \, AT \, (current = 1 \, A)$$

Magnetizing force H at a radius of 1 m is

$$H = \frac{IN}{\ell} = 1/(2\pi r) \, AT = 1/2\pi \qquad (r = 1)$$

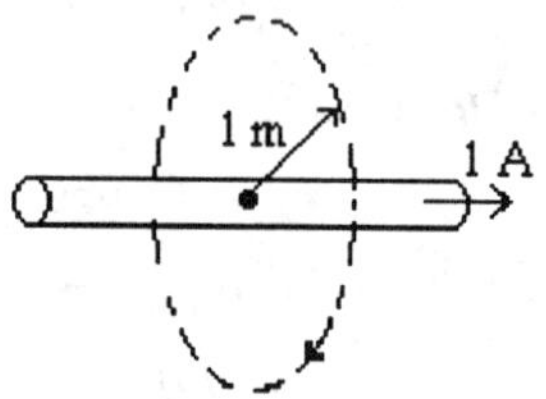

Fig. 3.15-1

In free space at a radius of 1 m, the ratio

$$\frac{B}{H} = \mu_o$$

$$B = \mu_o H = \mu_o \times 1/2\pi$$

$$= \mu_o/2\pi \, T \qquad (Eq. \ 3.15\text{-}0)$$

Now consider the second conductor which lies in the field of radius 1 m.
Force on this conductor, $F = 2 \times 10^{-7}$ N/m

$$But \qquad F = BI\ell \text{ newtons}$$

$$\therefore \qquad B = F/(I \times \ell) = 2 \times 10^{-7}/(1 \times 1)$$

$$= 2 \times 10^{-7} \, T \qquad (Eq. \ 3.15\text{-}1)$$

Equating Eq. 3.15-0 and Eq. 3.15-1, we get

$$\mu_o/2\pi = 2 \times 10^{-7}$$

$$\therefore \qquad \mu_o = 4\pi \times 10^{-7} \text{ H/m}$$

(contd)

(b)   Flux density, $B = \Phi/A$

$$= \frac{0.5 \times 10^{-3}}{5 \times 10^{-4}} = \underline{1.0\ T}$$

Magnetizing force for iron, $H_i = 500$ AT/m   (from data given)

Magnetizing force for gap, $H_g = B/\mu_o$

$$= 1/(4\pi \times 10^{-7}) = \underline{7.96 \times 10^5\ AT/m}$$

Total m.m.f $= H_i\,\ell\,\text{(iron)} + H_g\,\ell\,\text{(gaps)}$

$$= (500 \times 0.20) + (7.96 \times 10^5 \times 0.0001 \times 2)$$

$$= 100 + (79.6 \times 2) = \mathbf{\underline{259\ AT}}$$

Pull/pole/unit area $= B^2/2\mu_o$

$$= 1^2/(2 \times 4\pi \times 10^{-7})$$

$$= 10^7/(2 \times 12.56) = \underline{398.089 \times 10^3}$$

Effective pull (both faces) $= 2 \times 398.089 \times 10^3 \times \text{area}$

$$= 2 \times 398.089 \times 10^3 \times 5 \times 10^{-4}$$

$$= 398.089 \times 10^3 \times 10^{-3}$$

$$= \mathbf{\underline{398.1\ N}}$$

Load supported (lb) $= 398.1 \times 0.225 = \mathbf{\underline{89.6\ lbf}}$

Example 3.16

Determine the ampere-turns necessary to establish an attractive force of
500 kg in an electromagnetic brake which has a total contact area of
50 cm^2; the effective length of the magnetic circuit in the iron is 40 cm.
Assume an air gap of 2 mm and a relative permeability of 250.

Solution

The attractive force (or pull) between surfaces is

Force $=$ mass x acceleration

$$= 500 \times 9.81 \ \text{newtons}$$

Now attractive force $= B^2A/2\mu_o$

$\therefore \qquad B^2A/2\mu_o = 500 \times 9.81$

and $\qquad B^2A = 500 \times 9.81 \times 2 \times 4\pi \times 10^{-7}$

$$B^2 = \frac{123.2136 \times 10^{-4}}{A}$$

$$= \frac{123.2316 \times 10^{-4}}{50 \times 10^{-4}} = \underline{2.464\ T}$$

Hence, flux density, $B = \sqrt{(2.464)} = \underline{1.57\ T}$

(contd)

The magnetizing force (H) required to produce this flux density in the *iron* is obtained from the expression

$$H = {}^B\!/_{\mu_o\mu_r}$$

$$= \frac{1.57}{4\pi \times 10^{-7} \times 250} \qquad = \underline{5000 \text{ AT/m}}$$

The magnetizing force (H) required to establish the flux density (1.57 T) in the *air gap* is obtained from the expression

$$H = {}^B\!/_{\mu_o\mu_r}$$

$$= {}^{1.57}\!/(4\pi \times 10^{-7}) \qquad (\mu_r = 1 \text{ for air})$$

$$= {}^{1.57 \times 10^7}\!/_{4\pi} \qquad = \underline{1.25 \times 10^6 \text{ AT/m}}$$

Ampere-turns required for iron

Effective length (ℓ) of magnetic circuit in the *iron* is

$$\ell = 40 \text{ cm} = 0.4 \text{ m}$$

Ampere-turns (AT) for *iron* $= H \times \ell$

$$= 5000 \times 0.4 \qquad = \underline{2000 \text{ AT}}$$

Ampere-turns required for air gap

Length of *air gap*, $\ell = 2 \text{ mm} = 2 \times 10^{-3}$

Ampere-turns (AT) for *gap* $= H \times \ell$

$$= 1.25 \times 10^6 \times 2 \times 10^{-3}$$

$$= 2.50 \times 10^3 \qquad = \underline{2500 \text{ AT}}$$

Ampere-turns required for the whole circuit is

Total ampere-turns $= AT_{(iron)} + AT_{(gap)}$

$$= 2000 + 2500 \qquad = \underline{\mathbf{4500 \text{ AT}}}$$

Example 3.17

A cast iron ring having a mean circumference of 20 cm and circular cross-section 2 cm in diameter is wound with a coil of 80 turns as shown in Fig. 3.17-0 (winding partially shown). Calculate the current required to set up a flux of 0.2 mWb. Use Fig. 3.11-1 B/H curves.

Solution

$$\text{Cross-sectional area} = {}^{\pi d^2}\!/_4$$

$$= 3.14 \times {}^{2^2}\!/_4 \qquad = \underline{3.14 \text{ cm}^2}$$

(contd)

Flux density, B $= \Phi/A$

$$= \frac{0.2 \times 10^{-3}}{3.14 \times 10^{-4}} \qquad = \underline{0.637\ T}$$

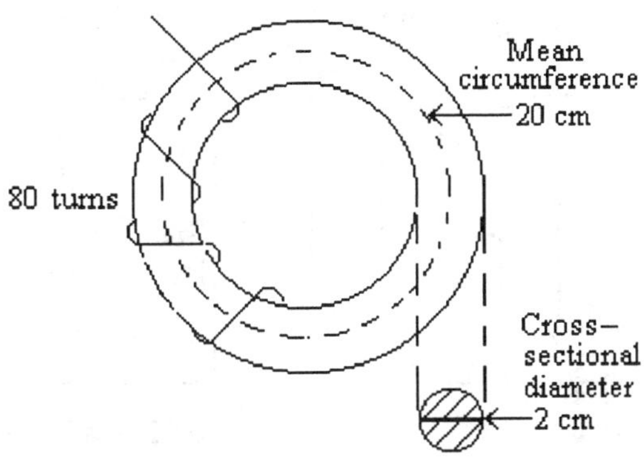

Fig. 3.17-0

From the B/H curves in Fig. 3.11-1 the corresponding magnetizing force
H for the flux density B (0.637 T) is

$$H = 3300\ AT/m\ \text{(approximately)}$$

Hence M.M.F required (AT) $= H \times \ell$ (ℓ = mean circumference)

$$= 3300 \times 20 \times 10^{-2} = \underline{660\ AT}$$

Now $\qquad\qquad$ M.M.F $= I\,N$

$\therefore \qquad\qquad\qquad I = M.M.F/N$

$$= {}^{660}/_{80} \qquad\qquad = \underline{\mathbf{8.25\ A}}$$

Example 3.18

(a) *State the modified form of Kirchhoff's Laws as applied to magnetic circuits.*

(b) *A magnetic circuit shown in Fig. 3.18-1 is constructed from Stalloy. The length of the flux path in the metal is 50 cm and the air gap is 4 mm. Calculate the value of the current required in a coil of 500 turns to produce a flux of 600 μWb. The cross-sectional area of the magnetic circuit is 5 cm². Use Fig. 3.11-1 B/H curves for Stalloy.*

Solution

(a) Kirchhoff's Laws for the electric circuit can also be applied to the magnetic circuit as follows:

(contd)

First Law

The total magnetic flux toward a junction is equal to the total magnetic flux away from that junction.

Consider Figs. 3.18-0(a) and (b)

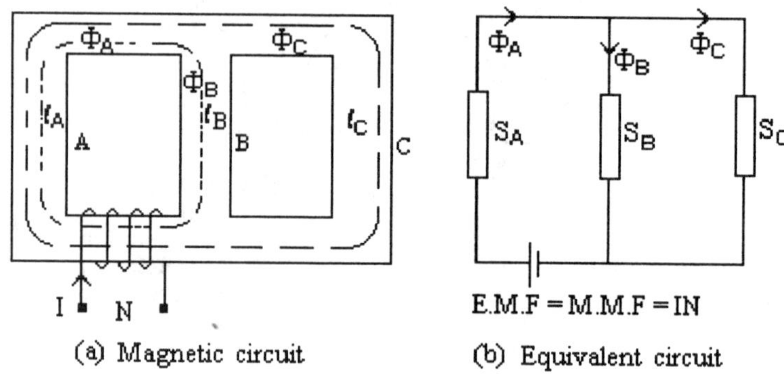

(a) Magnetic circuit   (b) Equivalent circuit

Fig. 3.18-0

Refer to Fig. 3.18-0(b)

Each line of magnetic flux forms a closed path such that

$$\Phi_A = \Phi_B + \Phi_C$$
$$\Phi_A - \Phi_B - \Phi_C = 0$$

In general $\quad \Sigma\Phi = 0$

Second Law

In any closed magnetic circuit the algebraic sum of the products of the magnetizing force (H) and the length (ℓ) of each path of the circuit is equal to the resultant magnetomotive force (m.m.f).

Refer to Fig. 3.18-0(a)

$$\text{M.M.F of coil} = I\,N$$
$$I\,N = H_A\,\ell_A + H_B\,\ell_B$$
$$= H_A\,\ell_A + H_C\,\ell_C$$
$$0 = H_B\,\ell_B - H_C\,\ell_C$$

In general $\quad \Sigma\,\text{M.M.F} = \Sigma H\ell$

(b)   Refer to Fig 3.18-1(b)

The reluctance of the iron part of the magnetic circuit is lumped as S_1.
The reluctance of the air gap is represented as S_2.

$$\text{Flux density, } B = \Phi/A = \frac{600 \times 10^{-6}}{5 \times 10^{-4}} = \underline{1.2\ \text{T}}$$

M.M.F for iron circuit

From Fig. 3.11-1 B/H curves, the magnetizing force required to produce this value of flux is

$$H = 560 \text{ turns} \quad \text{(approximately)}$$

$$\therefore \quad \text{M.M.F} = H \times \ell$$

$$= 560 \times 50 \times 10^{-2} \qquad = \underline{280 \text{ AT}}$$

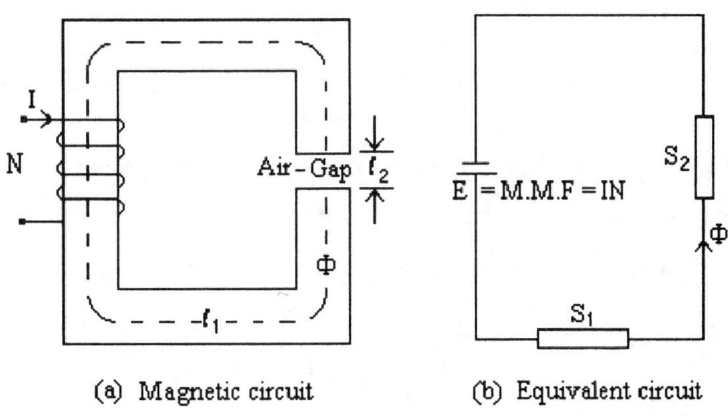

(a) Magnetic circuit   (b) Equivalent circuit

Fig. 3.18-1

M.M.F for air gap

The number of ampere-turns (AT) required to produce the flux in the air gap is

$$\text{M.M.F} = \Phi S_2 = \Phi \ell_2 / \mu_o A$$

$$= \frac{600 \times 10^{-6} \times 4 \times 10^{-3}}{4\pi \times 10^{-7} \times 50 \times 10^{-3}}$$

$$= \frac{600 \times 4 \times 10^{-9}}{4\pi \times 5 \times 10^{-11}}$$

$$= \frac{120 \times 10^2}{\pi} \qquad = \underline{3821.7 \text{ AT}}$$

Total M.M.F (AT) requirement for the complete circuit is

$$\text{M.M.F}_{(total)} = \text{M.M.F}_{(iron)} + \text{M.M.F}_{(air\text{-}gap)}$$

$$= 280 + 3821.7 \qquad = \underline{\mathbf{4102 \text{ AT}}}$$

Example 3.19

Derive an expression for the value of Φ_1 and Φ_2 in terms of the reluctance of the magnetic circuit shown in Fig. 3.19-0

(contd)

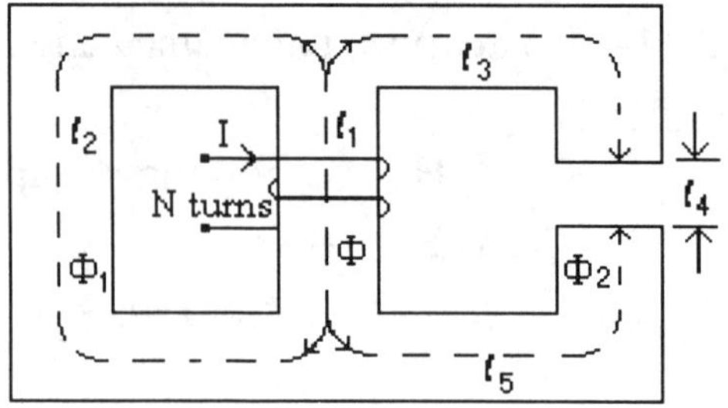

Fig. 3.19-0

Solution

The magnetic circuit shown in Fig. 3.19-0 is deemed a series-parallel magnetic circuit.

For simplicity, an equivalent circuit is shown in Fig. 3.19-1.

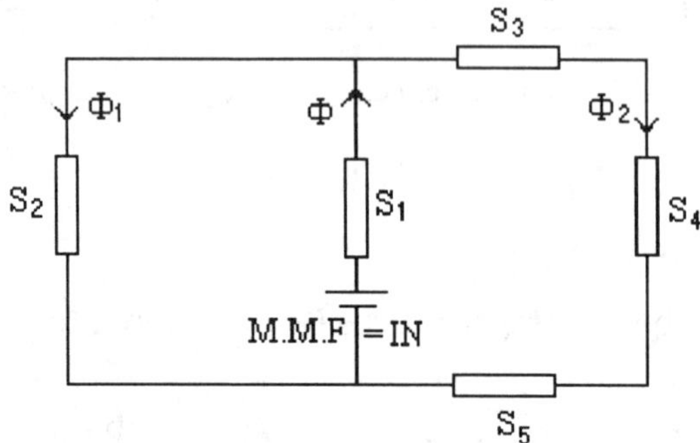

Fig. 3.19-1

Note:

Refer to Fig. 3.19-0

Observe that the right-hand outer limb is broken into 3 lengths:

ℓ_3, ℓ_4 (air-gap), and ℓ_5. As a result, it is represented by 3 separate reluctances in the equivalent circuit shown in Fig. 3.19-1.

Refer to Fig. 3.19-1

The values of the circuit elements may now be obtained from the equivalent circuit. Thus

$$\text{M.M.F} = I\,N$$
$$S_1 = \ell_1/\mu_o\mu_r\,A_1$$
$$S_2 = \ell_2/\mu_o\mu_r\,A_2$$
$$S_3 = \ell_3/\mu_o\mu_r\,A_3$$
$$S_4 = \ell_4/\mu_o\mu_r\,A_4$$
$$S_5 = \ell_5/\mu_o\mu_r\,A_5$$

(contd)

In the above, ℓ, μ_r and A represent the individual length, relative permeability and area, respectively, of the individual sections of the iron.

$$\text{Total reluctance, } S_T = S_1 + \frac{S_2(S_3 + S_4 + S_5)}{S_2 + S_3 + S_4 + S_5}$$

$$\text{Total flux, } \Phi = \text{M.M.F}/S_T$$

Expressions for Φ_1 and Φ_2 are as follows:

$$\Phi_1 = \frac{\Phi}{S_2 + S_3 + S_4 + S_5} \times (S_3 + S_4 + S_5)$$

$$= \frac{\Phi(S_3 + S_4 + S_5)}{S_2 + S_3 + S_4 + S_5}$$

$$\Phi_2 = \frac{\Phi}{S_2 + S_3 + S_4 + S_5} \times S_2$$

$$= \frac{\Phi S_2}{S_2 + S_3 + S_4 + S_5}$$

Alternatively

As an alternative, let us assume (for further simplicity) that the reluctances S_3, S_4 and S_5 are replaced by an equivalent reluctance Seq as shown in Fig. 3.19-2.

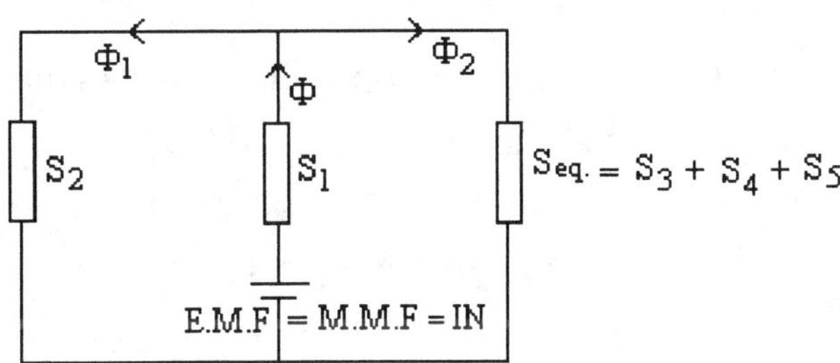

Fig. 3.19-2

$$S_{eq} = S_3 + S_4 + S_5$$

$$S_T = S_1 + \frac{S_2 \, S_{eq.}}{S_2 + S_{eq.}}$$

$$\therefore \quad \Phi_1 = \frac{\Phi \, S_{eq.}}{S_2 + S_{eq.}}$$

$$\text{and} \quad \Phi_2 = \frac{\Phi S_2}{S_2 + S_{eq.}}$$

Example 3.20

*A cast steel electromagnet has a magnetic circuit which may be regarded
as three parts in series, each of uniform cross-section.*
Part (a) has a length of 8 cm and cross-sectional area 0.5 cm².
Part (b) has a length of 6 cm and cross-sectional area 0.9 cm².
Part (c) has an air gap of length 0.5 mm and cross-sectional area 1.5 cm².
*Determine the current required in a coil of 4000 turns wound on part (b)
to produce in the air gap a flux density of 0.3 T. Magnetic leakage may be
neglected. Use the data from the B/H curves in Fig. 3.11 -1.*

Solution

To determine the flux Φ for the whole magnetic circuit, it is necessary to
calculate the flux Φ for the air gap. Hence

$$\text{Flux, } \Phi = B_{gap} \times A_{gap}$$
$$= 0.3 \times 1.5 \times 10^{-4}$$
$$= 3 \times 15 \times 10^{-6} \qquad = \underline{45\ \mu Wb}$$

Consider part (a)

$$\text{Flux density , B} = {}^{\Phi}/A$$
$$= \frac{45 \times 10^{-6}}{0.5 \times 10^{-4}} \qquad = \underline{0.9\ T}$$

From Fig. 3.11-1 B/H curves, the corresponding magnetizing force
for mild steel at this flux density (0.9 T) is

$$H = 875\ \text{AT/m}\quad\text{(approximately)}$$
$$\text{M.M.F required} = H \times \ell$$
$$= 875 \times 8 \times 10^{-2} \qquad = \underline{70\ AT}$$

Consider part (b)

$$\text{Flux density, B} = {}^{\Phi}/A$$
$$= \frac{45 \times 10^{-6}}{0.9 \times 10^{-4}} \qquad = \underline{0.5\ T}$$

From Fig. 3.11-1 B/H curves, the corresponding magnetizing force at this
flux density (0.5 T) is

$$H = 375\ \text{AT/m}\quad\text{(approximately)}$$
$$\text{M.M.F required} = H \times \ell$$
$$= 375 \times 6 \times 10^{-2} \qquad = \underline{22.5\ AT}$$

(contd)

Consider part (c) - air gap

$$H = B/\mu_o$$

and $$M.M.F = H \times \ell$$

$\therefore$ $$M.M.F = B\ell/\mu_o$$

$$= \frac{0.3 \times 0.5 \times 10^{-3}}{4\pi \times 10^{-7}}$$

$$= \frac{15 \times 10^2}{4\pi} \qquad = \underline{119.43\ AT}$$

Total M.M.F $= M.M.F_{(part\ a)} + M.M.F_{(part\ b)} + M.M.F_{(gap)}$

$$= 70 + 22.5 + 119.43 = \underline{211.93\ AT}$$

Now   M.M.F $= I\,N$

$\therefore$ Required current, I $= M.M.F/N$

$$= 211.93/4000 \qquad = \underline{\mathbf{0.05298\ A}}$$

$$\mathbf{or\ \underline{52.98\ mA}}$$

Example 3.21

A magnetic core made of cast steel has the dimensions shown in Fig. 3.21-0. The cross-sectional area of the outer limbs is 8 cm² and the magnetic fringing increases the effective cross-sectional area of the air gap to 10 cm². The center limb has a cross-sectional area of 6 cm² and is wound with a coil of 800 turns. The leakage coefficient is 1.15. Estimate the current required to establish a flux of 100 μWb in the air gap. Use Fig. 3.11-1 B/H curves.

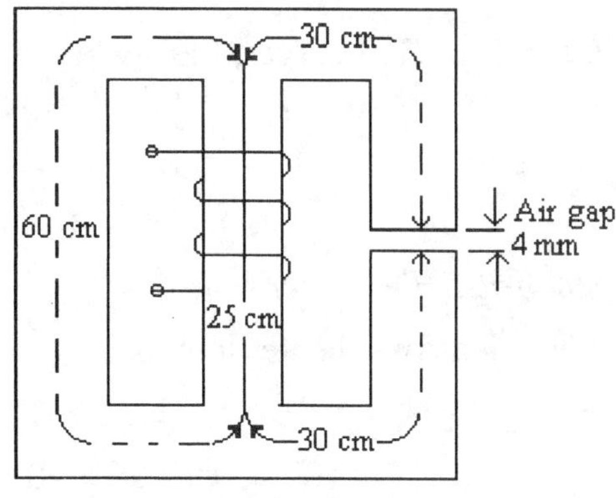

Fig. 3.21-0

(contd)

Solution

First, consider the air gap [100 µWb to be established].

$$\text{Flux density, B} = \Phi/A$$
$$= \frac{100 \times 10^{-6}}{10 \times 10^{-4}} = \underline{0.1 \text{ T}}$$

Now $\quad B/H = \mu_o\mu_r$

$\therefore\quad$ Magnetizing force, $H = B/\mu_o \quad (\mu_r \text{ for air} = 1)$

$$= 0.1/(4\pi \times 10^{-7}) = \underline{79618 \text{ AT/m}}$$
$$\text{M.M.F (AT)} = H \times \ell \quad (AT)$$
$$= 79618 \times 4 \times 10^{-3} = \underline{318.5 \text{ AT}}$$

Second, consider the right-outer limb.

$$B = \Phi/A$$
$$= \frac{100 \times 10^{-6}}{8 \times 10^{-4}} = \underline{0.125 \text{ T}}$$

From Fig. 3.11-1 B/H curves for cast steel

$$H = 125 \text{ AT/m (approximately)}$$
$$\text{M.M.F} = H \times \ell \quad (AT)$$
$$= 125 \times (30 \times 2) \times 10^{-2} = \underline{75 \text{ AT}}$$

Summarizing:

$$\text{Total m.m.f} = \textit{right-outer limb m.m.f} + \textit{air gap m.m.f.}$$
$$\text{Total m.m.f} = 75 + 318.5 = \underline{393.5 \text{ AT}}$$

Now this m.m.f (393.5 AT) is also applied to the left-outer limb.

$\therefore$ *Third, considering the left-outer limb*

$$H = \text{m.m.f}/\ell$$
$$= 393.5/(60 \times 10^{-2}) = \underline{655.8 \text{ AT/m}}$$

From Fig. 3.11-1 B/H curves, the analogous flux density for cast steel is

$$B = 0.7 \text{ T (approximately)}$$
$$\therefore\quad \Phi = B \times A$$
$$= 0.7 \times 8 \times 10^{-4} = \underline{0.56 \text{ mWb}}$$

Finally, consider the center limb.

Flux density in center limb is

$$\Phi = (\Phi_{limb} + \Phi_{gap}) \times \text{leakage coefficient}$$
$$= (0.56 \times 10^{-3} + 100 \times 10^{-6}) \times 1.15$$
$$= (0.56 \times 10^{-3} + 0.1 \times 10^{-3}) \times 1.15$$
$$= (0.66 \times 10^{-3}) \times 1.15 = \underline{0.759 \times 10^{-3} \text{ Wb}}$$

(contd)

$$\text{Flux density, } B = \Phi/A$$

$$= \frac{0.759 \times 10^{-3}}{6 \times 10^{-4}} \qquad = \underline{1.265 \text{ T}}$$

From Fig. 3.11-1 B/H curves, the corresponding value of magnetizing force

$$H = 1625 \text{ AT/m (approximately)}$$

$$M.M.F = H \times \ell$$

$$= 1625 \times 25 \times 10^{-2} = \underline{406.25 \text{ AT}}$$

$$M.M.F \text{ for the whole circuit} = m.m.f_{\text{(outer limb and gap)}} + m.m.f_{\text{(center limb)}}$$

$$= 393.5 + 406.25 \approx \underline{800 \text{ AT}}$$

$$\text{Current required, } I = M.M.F/N$$

$$= {}^{800}/_{800} \qquad = \underline{\textbf{1.0 A}}$$

Example 3.22

The arm of a d.c motor starter is held in the "ON" position by means of an electromagnet as shown in Fig. 3.22-0. The spiral spring exerts a mean counter torque of 2 lbf-ft on the armature. The length between the center of the armature and the pivot on the starter arm is 6 in. The cross-sectional area of each pole face of the electromagnet is 3.5 cm^2 and the air gap is 0.5 mm. Estimate the minimum number of turns required to keep the arm in the "ON" position against the force of the spring. Neglect the reluctance of the iron path and the effects of magnetic leakage and fringing. [Take 1 N = 0.225 lb]

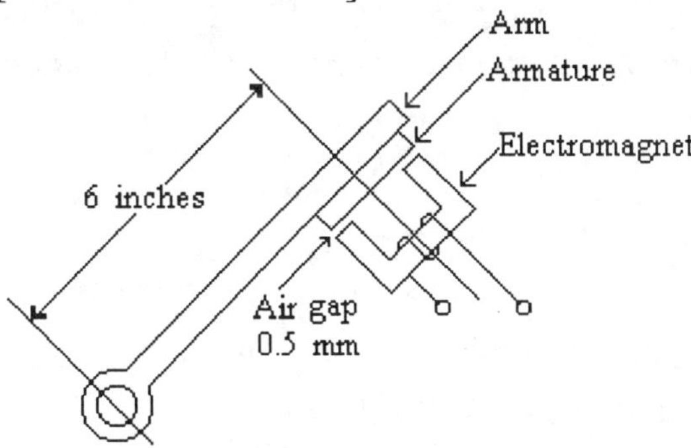

Fig. 3.22-0. Arm of a d.c motor starter

Solution

$$\text{Torque (T) at armature} = \text{force (F) x distance (d)}$$

$$\therefore \qquad \text{Force, } F = {}^{T}/_{d} = {}^{2}/_{0.5} \qquad = \underline{4 \text{ lb}}$$

$$\text{Pull per pole} = B^2 A/2\mu_o \qquad \text{newtons (N)}$$

(contd)

$$\text{Pull per pole in newtons} = 2 \text{ lb} = {}^2/0.225 \text{ N}$$

$$\text{Now} \quad B^2A/2\mu_o = {}^2/0.225$$

$$B^2 = \frac{2 \times 2\mu_o}{0.225 \times A}$$

$$B^2 = \frac{2 \times 2 \times 4\pi \times 10^{-7}}{0.225 \times 3.5 \times 10^{-4}}$$

$$= \frac{16\,\pi \times 10^{-7}}{0.7875 \times 10^{-4}}$$

$$= \frac{16\,\pi \times 10^{-3}}{0.7875}$$

$$= 20.32\,\pi \times 10^{-3} \qquad = \underline{0.0638}$$

$$\therefore \qquad B = \sqrt{(0.0638)} \qquad = \underline{0.253 \text{ T}}$$

$$\text{Magnetizing force, H} = B/\mu_o$$

$$= 0.253/(4\pi \times 10^{-7}) \quad = \underline{201433 \text{ AT/m}}$$

$$\text{Minimum m.m.f} = H \times \ell \qquad (\ell = 2 \text{ gaps} = 2 \times 0.5 \text{ mm})$$

$$= 201433 \times (2 \times 0.5) \times 10^{-3}$$

$$= 201433 \times 1.0 \times 10^{-3}$$

$$= 201443 \times 0.001 \qquad = \underline{\mathbf{201.4 \text{ AT}}}$$

Example 3.23

(a)   *Define the unit of self-inductance and deduce the expression for the average e.m.f of self-inductance.*

(b)   *Calculate the average value of the self-induced e.m.f in a coil having an inductance of 0.5 H when the current is increased from 0.2 A to 2.2 A in 0.05 sec.*

Solution

(a)   A circuit has an inductance of one henry (1 H) if an e.m.f of one volt (1 V) is induced in the circuit when the current changes uniformly at the rate of one ampere (1 A) per second.

The induced e.m.f is doubled if the inductance or change in current is doubled.

Induced e.m.f accompanied by increase or decrease in current

If the current in a circuit increases from I_1 to I_2 amperes in t sec, then

$$\text{Average rate of change of current} = (I_2 - I_1)/t$$

(contd)

$$\text{Average induced e.m.f, } e = - L \times \text{rate of change of current}$$

$$= - L \, di/dt \text{ volts}$$

The minus sign indicates that the induced e.m.f e is in opposition to the applied voltage.

(b)

$$\text{Average induced e.m.f, } e = - L \, di/dt$$

$$= - 0.5 \times (I_2 - I_1)/t$$

$$= - 0.5 \times \frac{(2.2 - 0.2)}{0.05} = - \underline{\mathbf{20\ V}}$$

Example 3.24

A coil of 150 turns wound on an iron core has an inductance of 0.5 H. Calculate

    (a)   the flux produced in the core when a current of 5 A is flowing in the coil.

    (b)   the value of the induced e.m.f in the coil when the current changes from +5 A to –5 A in 0.2 sec

Briefly derive any formula used.

Solution

Induced e.m.f accompanied by increase or decrease in flux, Φ

From the laws of magnetism, when the flux linkages in a coil of N turns are changing, the e.m.f induced in the coil is given by

$$e = - N \, d\Phi/dt \text{ volts}$$

This is the instantaneous value of the induced e.m.f. The average value of the induced e.m.f may be calculated as follows:

Assume that flux Φ_1 is the original flux linking with the coil of N turns at time t_1 and its value changes to Φ_2 at time t_2. Then

$$\text{Change of flux, } d\Phi = \Phi_2 - \Phi_1$$

$$\text{Time interval, } dt = t_2 - t_1$$

$$\therefore \text{ Average value of induced e.m.f, } e = - N \, d\Phi/dt$$

Also,   average value of induced e.m.f, $e = - L \, di/dt$

Thus

$$- L \, di/dt = - N d\Phi/dt$$

$$L = N \, d\Phi/di$$

$$= N \times \frac{\underline{\text{change in flux linkages}}}{\text{change in current}}$$

127

(contd)

$$\therefore \qquad L = N\Phi/I$$

$$\text{Now} \qquad \Phi = B\,A$$
$$= \mu_o\mu_r\,H\,A$$
$$= \mu_o\mu_r\,A\,N\,I/\ell \qquad [H = N\,I/\ell\,]$$

Substituting for Φ,

$$L = \mu_o\mu_r\,A\,N^2/\ell \qquad \text{(Eq. 3.24-0)}$$

[Note: From Eq. 3.24-0, the inductance of a coil (L) is proportional to N^2. Hence, doubling the number of turns in the coil will increase its inductance 4 times.]

(a) Flux Φ produced is obtained as follows:

$$L = N(\Phi_2 - \Phi_1)/(I_2 - I_1)$$
$$= N\,\Phi_2/I_2 \qquad [\Phi_1 \text{ and } I_1 = 0]$$
$$\therefore \qquad \Phi = L\,I_2/N$$
$$= \frac{0.5 \times 5}{1500} \qquad = \mathbf{1.67\ mWb}.$$

(b) Since the current changes from +5 A to –5 A in 0.2 sec

$$\therefore \ \text{Change of current} = I_2 - I_1$$
$$= -5 - (+5) \qquad = -10\ A$$
$$\therefore \qquad \text{Induced e.m.f} = -L\,di/dt$$
$$= -0.5 \times (-10/0.2) \qquad = \mathbf{25\ V}$$

In this case the induced e.m.f is in the same direction as the supply voltage. Thus the induced e.m.f opposes the reversal of the current. The direction of the induced e.m.f always opposes the source producing it (Lenz's Law). The source in this case is the current.

Example 3.25

(a) *The circuit of Fig. 3.25-0(a) is a simple L-R circuit connected to a d.c supply via a single-pole switch. Explain with the aid of diagrams the circuit behavior when switch S is closed.*

(b) *Derive expressions for*

(i) *the rise of current in an L-R (inductive) circuit*

(ii) *the decay of current in an L-R (inductive) circuit*

(contd)

(c)   *A coil having a self-inductance of 2 H and a d.c resistance of 200 Ω is connected across a 100 V d.c supply of negligible internal resistance. Calculate the value of the current 0.002 sec after the switch is closed. Refer to Fig. 3.25-0(b).*

Solution

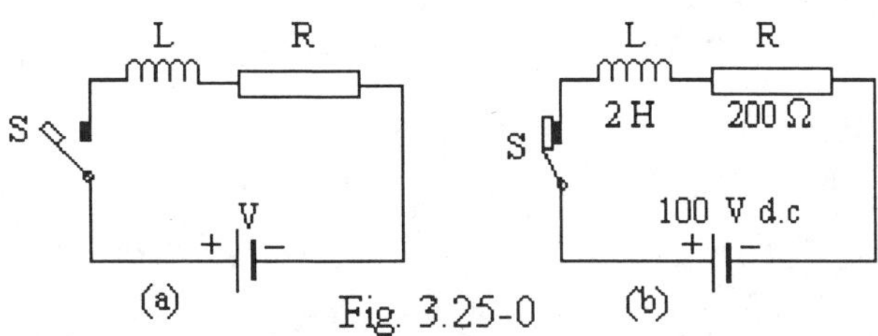

Fig. 3.25-0

(a)   <u>Consider Fig. 3.25-0(a)</u>

At the instant of closing switch S, the current does not immediately rise to its final steady value, $I = {}^V/R$. Rather the current rises exponentially.

At any instant of time, $V = iR + L\,{}^{di}/dt$

    where t = time after closing switch

       i = instantaneous current

$\therefore$ $V - L\,{}^{di}/dt = iR$

    At the instant of closing switch S

       $i = 0$

$\therefore$ $iR = 0$

and $V = L\,{}^{di}/dt$ or ${}^{di}/dt = {}^V/L$

This is the initial rate of change of current.

When the current I has reached its final steady value

    ${}^{di}/dt = 0$ and $V = IR$

Hence the final value of current is

    $I = {}^V/R$

(b)   **Rise and Decay of current in an L-R (inductive) circuit.**

<u>Refer to Fig. 3.25-1</u>

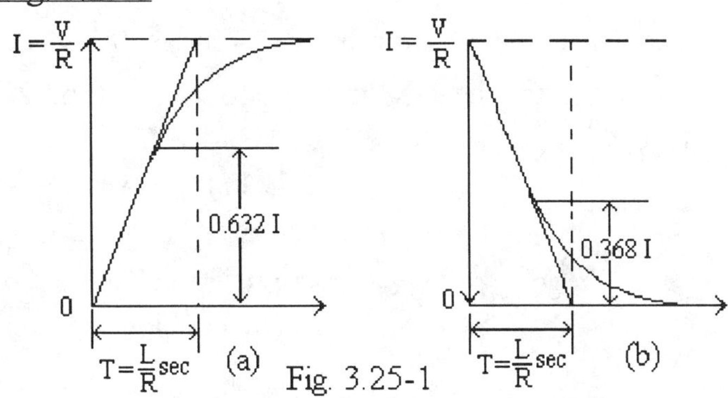

Fig. 3.25-1

129

(contd)

The rise of current in an L-R circuit is shown in Fig. 3.25-1(a).

The decay of current in an L-R circuit is indicated in Fig. 3.25-1(b).

(i)   *Rise of current in an L-R circuit*

<u>Refer to Fig. 3.25-1(a)</u>

At any instant of time, V $= i\mathrm{R} + \mathrm{L}\,di/dt$

Dividing by R, we obtain

$$V/\mathrm{R} = i + L/\mathrm{R}\,di/dt$$

That is, $\quad \mathrm{I} = i + L/\mathrm{R}\,di/dt$

and $\quad L/\mathrm{R}\,di/dt = \mathrm{I} - i$

Inverting, $\quad \mathrm{R}/L\,dt/di = 1/(\mathrm{I} - i)$

Integrating both sides, we get

$$\int (\mathrm{R}/L)\,dt/di = \int 1/(\mathrm{I} - 1)$$

Let $\quad \mathrm{Z} = (\mathrm{I} - i);\ $ then $di/dz = -1$

$$\therefore \quad \int (\mathrm{R}/L)\,dt/di = \int 1/\mathrm{z}$$

$$\int (\mathrm{R}/L)\,dt = \int di/\mathrm{z}$$

$$= \int 1/\mathrm{z}\,.\,di/dz\,.\,d\mathrm{Z}$$

$$\mathrm{R}t/L = \log_e \mathrm{Z} \times (-1) + \mathrm{C}$$

$$\mathrm{R}t/L = -\log_e (\mathrm{I} - i) + \mathrm{C} \qquad \text{(Eq. 3.25-0)}$$

When $\quad t = 0,\quad i \quad = 0$

$$\therefore \quad 0 = -\log_e \mathrm{I} + \mathrm{C}$$

and $\quad \mathrm{C} = \log_e \mathrm{I}$

$$\therefore \quad \mathrm{R}t/L = -\log_e (\mathrm{I} - i) + \log_e \mathrm{I}$$

$$-\mathrm{R}t/L = \log_e[(\mathrm{I} - i)/\mathrm{I}]$$

$$\therefore \quad e^{-\mathrm{R}t/L} = (\mathrm{I} - i)/\mathrm{I}$$

$$\mathrm{I}e^{-\mathrm{R}t/L} = \mathrm{I} - i$$

$$i = \mathrm{I} - \mathrm{I}e^{-\mathrm{R}t/L}$$

$$\therefore \quad i = \mathbf{I\,(1 - e^{-Rt/L})} \qquad \text{(Eq. 3.25-1)}$$

The *time constant* T for an L-R circuit is L/R seconds. It is defined as the time taken for the current to rise to 0.632 (or 63.2%) of its final value. Hence

$$i = \mathrm{I}\,(1 - e^{-\mathrm{R}/L \times L/\mathrm{R}}) \quad (t = L/\mathrm{R})$$

$$\therefore \quad i = \mathbf{I\,(1 - e^{-1})} = \mathbf{0.632\,I} \qquad \text{(Eq. 3.25-2)}$$

$[\,e = 2.7183\ $ and $\ 1/e \approx 0.36788\,]$

(contd)

(ii)   *Decay of current in an L-R circuit*

<u>Refer to Fig. 3.25-1(b)</u>

From Eq. 3.25-0, determine the value of C.

That is, $\quad\quad Rt/L = -\log_e (I - i) + C$

$\quad\quad$ When $t = 0$, $\quad i = V/R$

$\therefore \quad$ Final value of $\quad I = 0$

Substituting these values in Eq. 3.25-0, we get

$$0 = -\log_e (0 - V/R) + C$$

from which $\quad C = \log_e(- V/R)$

Therefore, $\quad Rt/L = -\log_e(I - i) + \log_e (-V/R)$

But the final value of $I = 0$

Therefore, $\quad Rt/L = -\log_e(- i) + \log_e(- V/R)$

$$= \log_e [i/(V/R)]$$

and $\quad\quad e^{-Rt/L} = i/(V/R)$

Therefore, $\quad i = Ie^{-Rt/L}$ $\quad\quad\quad\quad$ (Eq. 3.25-3)

In the case of the decay of current, the time constant $(T = L/R \text{ sec})$ is the time taken for the current to decay or fall to 0.368 (or 36.8%) from its final value. Thus

$$i = Ie^{-Rt/L}$$

$$\therefore \quad i = Ie^{-1} \quad\quad = 0.368\ I$$

$[e = 2.7183$ and $1/e \approx 0.36788]$

Note: Eq. 3.25-1, $i = I(1 - e^{-Rt/L})$, is referred to as the HELMHOLTZ EQUATION.

(c)   From Eq. 3.25-1

$$i = I (1 - e^{-Rt/L})$$

$$L = 2\ H$$

$$R = 200\ \Omega$$

$$I = V/R = 100/200 \quad = 0.5\ A$$

$$t = 0.002\ sec$$

$$\therefore \quad i = 0.5 (1 - e^{-200 \times 0.002/2})\ A$$

$$= 0.5 (1 - e^{-0.4/2})$$

$$= 0.5 (1 - e^{-0.2})$$

$$= 0.5 (1 - 0.8187)$$

$$= 0.5 \times 0.1813 \quad\quad = \underline{\textbf{0.09065 A}}$$

$$\underline{\textbf{or\ 90.65 mA}}$$

Example 3.26

A relay has a coil of resistance 25Ω and inductance 0.5 H. The relay is energized by a d.c voltage pulse which rises from 0 to 10 V instantly, remains constant for 0.25 sec and then falls immediately to zero. If the relay contacts close when the current reaches 200 mA (increasing) and open when the current falls to 100 mA (decreasing), determine for how long the contacts remain closed.

Solution

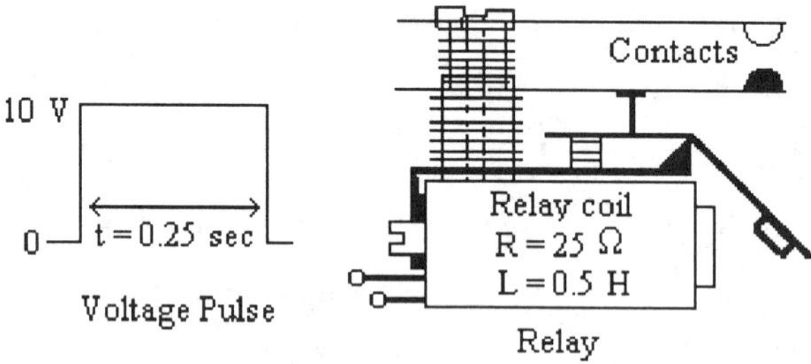

Fig. 3.26-0

Refer to Fig. 3.26-0, which is a visual aid of the relay with the d.c voltage pulse.

Time constant of relay coil is

$$T \quad = {}^{L}\!/_{R} = {}^{0.5}\!/_{25} \quad = \underline{0.02 \text{ sec}}$$

The current in the relay coil is

$$I \quad = {}^{V}\!/_{R} = {}^{10}\!/_{25} \quad = \underline{0.4 \text{ A}}$$

The voltage pulse remains constant at 10 V for 0.25 sec. This is longer than the time constant, which is 0.02 sec. Consequently, the relay current will practically attain its steady value long before the voltage pulse ends. The rise of current in the relay coil is given by

$$i \quad = I\,(1 - e^{-Rt/L}) \qquad \text{(refer to Eq. 3.25-1)}$$

where i = 200 mA (0.2 A), I = 0.4 A, e = 2.7183, R = 25 Ω, and L = 0.5 H.

$$\therefore \qquad 0.2 \quad = 0.4\,(1 - e^{-25 \times t/0.5})$$
$$= 0.4\,(1 - e^{-50t})$$
$${}^{0.2}\!/_{0.4} \quad = 1 - e^{-50t}$$
$$e^{-50t} \quad = 1 - {}^{2}\!/_{4}$$
$$= {}^{1}\!/_{2}$$

Inverting, $\qquad e^{50t} \quad = 2$

Taking logarithm to base e of both sides

(contd)

$$50t \log_e e = \log_e 2$$
$$50t \times 1 = 0.6931$$
$$\therefore \quad t = {}^{0.6931}/_{50} = \underline{0.01386 \text{ sec}}$$

Alternatively, taking logarithm to base 10 of both sides

$$50t \log_{10} e = \log_{10} 2$$
$$50t \times \log_{10} 2.7183 = \log_{10} 2$$
$$50t \times 0.4343 = 0.3010$$
$$\therefore \quad t = {}^{0.3010}/(50 \times 0.4343)$$
$$= {}^{0.3010}/21.715 = \underline{0.01386 \text{ sec}}$$

The relay contacts close at $t = 0.01386$ sec after the pulse is applied at $t = 0$.

At the end of the voltage pulse (i.e., after 0.25 sec), the current in the relay falls according to Eq. 3.25-3. Thus

$$i = Ie^{-Rt/L}$$

where $i = 100$ mA $= 0.1$ A and

$$\text{I, R and L are unchanged.}$$

$$\therefore \quad 0.1 = 0.4 \, e^{-50t}$$
$$e^{-50t} = {}^{0.1}/0.4$$

Inverting,
$$e^{50t} = {}^{4}/1$$
$$50t \log_{10} e = \log_{10} 4$$
$$50t \log_{10} 2.7183 = \log_{10} 4$$
$$50t \times 0.4343 = 0.6021$$
$$\therefore \quad t = {}^{0.6021}/(50 \times 0.4343)$$
$$= {}^{0.6021}/21.715 = \underline{0.02773 \text{ sec}}$$

The time period for which the relay contacts were closed is as follows:

$$\text{Pulse time} = 0.25 \text{ sec} \qquad (A)$$

Time at which contacts were initially closed while current was increasing

$$= 0.01386 \text{ sec} \qquad (B)$$

Time at which contacts were opened while current was decreasing

$$= 0.02773 \text{ sec} \qquad (C)$$

Time duration in which contacts remained closed is

$$= A - (B + C)$$
$$= 0.25 - (0.01386 + 0.02773) \text{ sec}$$
$$= 0.25 - 0.04159 = \underline{\textbf{0.20841 sec}}$$
$$\underline{\textbf{or 208.41 msec}}$$

Example 3.27

A non-magnetic ring 5 cm in diameter and 100 cm long is uniformly wound with 2000 turns in a single layer. A secondary winding of 1000 turns is uniformly wound over the first winding near its center. Calculate

> *(a)   the self-inductance of each winding*
>
> *(b)   the mutual inductance between the two coils*
>
> *(c)   the total inductance obtained by connecting the two windings in*
>
> > *(i) series aiding (cumulatively coupled)*
> >
> > *(ii) series opposing (differentially coupled)*

Solution

(a)   Let L_{pry} = inductance of first coil

$\therefore$ $L_{pry} = N^2 \mu_o \mu_r \, A_{/\ell}$ H $(\mu_r = 1)$

Area, $A = \pi d^2 / 4$

$$= \frac{\pi \times 5^2 \times 10^{-4}}{4}$$

Length, $\ell = 100 \text{ cm} = 100 \times 10^{-2} \text{ meter}$

$\therefore$ $L_{pry} = \dfrac{2000^2 \times 4\pi \times 10^{-7} \times \pi \times 25 \times 10^{-4}}{100 \times 10^{-2} \times 4}$ H

$$= \frac{4 \times 10^6 \times 4\pi^2 \times 10^{-7} \times 25 \times 10^{-4}}{100 \times 10^{-2} \times 4} \quad \text{H}$$

$$= \frac{4 \times 4\pi^2 \times 25 \times 10^{-5}}{100 \times 10^{-2} \times 4}$$

$$= \pi^2 \times 10^{-3} \qquad = \underline{\mathbf{9.86 \ mH}}$$

Now from (1) above, $L_{pry} \propto N^2$

Consequently, doubling the number of turns will increase the inductance 4 times.

Further, the turns in the primary (2000) is twice the number of turns in the secondary (1000).

$\therefore$ $L_{pry} = 4 \times L_{sdy}$

and $L_{sdy} = \tfrac{1}{4} \times L_{pry}$

$$= \tfrac{1}{4} \times 9.86 \qquad = \underline{\mathbf{2.46 \ mH}}$$

Alternatively

$L_{sdy} = N^2 \mu_o \mu_r \, A_{/\ell}$ H

(contd)

$$= \frac{1000^2 \times 4\pi \times 10^{-7} \times \pi \times 25 \times 10^{-4}}{100 \times 10^{-2} \times 4}$$

$$= 1000^2 \times \pi^2/4 \times 10^{-9}$$

$$= 10^6 \times \pi^2/4 \times 10^{-9}$$

$$= \pi^2/4 \times 10^{-3} \qquad\qquad = \underline{\textbf{2.46 mH}}$$

(b) Mutual inductance, $M = \dfrac{\text{change of flux linkage with secondary}}{\text{change of current in primary}}$

Let Φ_1 = flux in the primary winding

i_1 = current in the primary winding

Then $\qquad M = N_{sdy}\, d\Phi_1/di_1 \qquad\qquad$ (Eq. 3.27-0)

Now $\qquad L_{pry} = N_{pry} \times d\Phi_1/di_1$

$\therefore \qquad d\Phi_1/di_1 = L_{pry}/N_{pry}$

Substituting for $d\Phi_1/di_1$ in Eq. 3.27-0, we get

$$M = N_{sdy} \times L_{pry}/N_{pry}$$

$$= 1000 \times \frac{9.86 \times 10^{-3}}{2000} \qquad\qquad = \underline{\textbf{4.93 mH}}$$

(c) *Total inductance:*

(i) Series aiding (cumulative) $= L_1 + L_2 + 2M$

$$= 9.86 + 2.46 + (2 \times 4.93)$$

$$= 9.86 + 2.46 + 9.86 \qquad\qquad = \underline{\textbf{22.18 mH}}$$

(ii) Series opposing (differential) $= L_1 + L_2 - 2M$

$$= 9.86 + 2.46 - 9.86 \qquad\qquad = \underline{\textbf{2.46 mH}}$$

Example 3.28

The area of the hysteresis loop for a certain magnetic material is 25 cm^2
and the scales (coordinates) of the graph upon which it is drawn are

$$1\ cm = 500\ AT/m$$

$$1\ cm = 0.3\ T$$

Calculate the hysteresis loss in watts per kilogram. Take the frequency
as 60 Hz and the density of the material as 8.5 g/cm^3.

Solution

The area of the B-H loop represents the energy dissipated in joules per
cubic meter.

(contd)

Total hysteresis loss/cycle in joules

$$= [\text{area of loop in cm}^2] \times [\text{amperes/cm}$$
$$(H) \text{ of scale}] \times [\text{webers/cm (B) of scale}]$$

Hysteresis loss/cycle $\quad = $ area of loop $\times$ H (AT/m) $\times$ B (T)

$$= 25 \times 500 \times 0.3 \quad = \underline{3750 \text{ J/m}^3}$$

Hysteresis loss for 60 Hz $\quad = 3750 \times 60 \quad = \underline{225 \text{ kJ/m}^3}$

This is the number of joules/second dissipated, or power loss, in the material in watts.

$$\therefore \quad \text{Power loss} \quad = \underline{225 \text{ kW/cubic meter (m}^3)}$$

The power loss per kilogram is as shown below.

$$1 \text{ cm}^3 \quad = 8.5 \text{ g}$$
$$1 \text{ m}^3 \quad = 8.5 \times 10^6 \text{ g}$$
$$\therefore \quad 1 \text{ kg of material} = 10^3/(8.5 \times 10^6) \quad \text{m}^3$$

Power loss in 1 m^3 of material $= 225 \times 10^3$ W

Power loss in 1 kg of material $= 225 \times 10^3 \times \dfrac{10^3}{8.5 \times 10^6} \quad = \underline{\mathbf{26.47 \ W/kg}}$

Example 3.29

A transformer has a hysteresis loss of 1800 W and an eddy current loss of 370 W at 30 Hz when the core flux density is 1.2 T. Estimate the total iron loss at 50 Hz if the maximum core flux density is then 1.0 T. [Take the Steinmetz Index as 1.6]

Solution

The losses occurring in ferromagnetic materials under the influence of alternating current magnetization are

      (i)   hysteresis loss

and   (ii) eddy current loss

(i)   Hysteresis loss $\propto$ f (B$_{max}$)^n [n = Steinmetz index]

(ii) Eddy current loss $\propto$ f^2 (B$_{max}$)^2

At 50 Hz

Hysteresis loss $= 1800 \times {}^{50}/30 \times [{}^{1.0}/1.2]^{1.6}$

$$= 1800 \times 1.67 \times 0.74698 \quad = \underline{2245.4 \text{ W}}$$

(contd)

$$\text{Eddy current loss} = 370 \times [{}^{50}/30]^2 \times [{}^{1.0}/1.2]^2$$
$$= 370 \times 1.67^2 \times 0.6944$$
$$= 370 \times 2.7889 \times 0.6944 \qquad = \underline{716.5 \text{ W}}$$
$$\text{Total iron loss} = \text{hysteresis loss} + \text{eddy current loss}$$
$$= 2245.4 + 716.5 \qquad = \underline{\mathbf{2961.9\ W}}$$

Example 3.30

Two coils A and B are magnetically coupled and a fluxmeter monitors the variation of flux in B. When a current of 5.0 A in coil A is reversed instantaneously, the deflection on the monitor is 80 divisions. The constant of the fluxmeter is 0.2 mWb/division.

Calculate

    (a)   the mutual inductance between the coils

    (b)   the coefficient of coupling if the self-inductances of coils A and B are 16 mH and 4 mH, respectively

Solution

(a)

$$\text{Mutual Inductance,}\quad M = \frac{\text{change of flux linkage in B}}{\text{change of current in A}}$$

Since the current is reversed instantaneously

$$\text{Change of current} \quad = 2 \times 5 \qquad = 10 \text{ A}$$

$$\therefore \qquad M = \frac{80 \times 0.2 \times 10^{-3}}{10} \qquad = \underline{1.6 \text{ mH}}$$

(b)   Refer to the introduction (Nomenclature, Formulae and Symbols)

Coefficient of coupling, $\quad k = {}^{M}/\sqrt{(L_1 L_2)}$

It will not be out of place here to briefly elaborate on the coefficient of coupling k.

Mutual inductance, $\quad M \qquad = \sqrt{(L_1 L_2)}$

This equation assumed that

(i)   the reluctance S is constant and

(ii)   the magnetic leakage is zero, that is, 100% of the magnetic flux produced by one coil links the other coil.

(contd)

In (i) above, the reluctance S is constant only when the magnetic circuit is of a non-magnetic medium (for example, air or vacuum).

In a magnetic medium the reluctance S is approximately constant. It may differ in magnetic circuits comprising different iron cores. Consequently, there will be magnetic leakage, and as a result less than 100% of the magnetic flux will link the other coil.

Hence

$$\text{Mutual inductance, } M = k \sqrt{(L_1 L_2)}$$

where k is termed the coefficient of coupling.

$$k = \frac{M}{\sqrt{(L_1 L_2)}}$$

$$= \frac{1.6 \times 10^{-3}}{\sqrt{(16 \times 4)} \times 10^{-3}}$$

$$= {}^{1.6}\!/_{8} \qquad = \underline{\mathbf{0.2}}$$

PROBLEMS   3

1.   The flux linking an air-cored coil of 500 turns changes from 30 μWb to
     60 μWb in 2 msec. Calculate the magnitude of the e.m.f induced in the coil.
     [*Ans. 7.5 V*]

2.   A conductor 100 cm long moves at a velocity of 25 m/sec in a direction at
     right angles to a uniform magnetic field of density 0.9 T. Calculate the
     value of the e.m.f induced in the conductor.   [*Ans. 22.5 V*]

3.   A conductor 50 cm long is moving at a velocity of 100 m/sec at an angle
     of 30° to a magnetic field of uniform flux density 0.4 T. Determine the
     value of the e.m.f developed in the conductor.   [*Ans. 10 V*]

4.   A conductor carries a current of 500 A at right angles to a magnetic field
     of uniform flux density 0.8 T. Calculate the force developed in the
     conductor in newtons per meter of length.   [*Ans. 400 N*]

5.   Determine the current which is required to set up a force of 10 lbf on a
     single conductor 50 cm long lying at right angles to a magnetic field
     of uniform flux density 0.9 T. (Take 1 N = 0.225 lbf)   [*Ans. 98.77 A*]

6.   A conductor of length 80 cm carries a current of 25 A in a direction at
     right angles to a uniform magnetic field of 0.9 T. Calculate
     (a)   the force in newtons on the conductor
     (b)   the power in watts required to move the conductor against the
             above force at a uniform speed of 10 m/sec.
                                      [*Ans. (a) 18 N, (b) 180 W*]

7.   Calculate the e.m.f generated in the axle of a motor vehicle traveling at
     90 km/h, assuming the length of the axle is 2 m and the vertical component
     of the earth's magnetic field is 40 μT.   [*Ans. 2 mV*]

8.   The armature of a four-pole generator having a magnetic flux of
     0.024 Wb/pole is driven at 1800 rev/min. Calculate the average e.m.f
     generated in one conductor of the armature.   [*Ans. 0.288 V*]

9.   A current increasing at a uniform rate from 0 to 20 A in 5 sec induces an
     e.m.f of 10 V in a coil. Calculate the self-inductance of the coil.
                                      [*Ans. 2.5 H*]

10.   Determine the e.m.f induced in a coil of inductance 0.4 H when the current
     in it is increasing at 150 A/sec.   [*Ans. 60 V*]

11.   A coil consisting of 500 turns has a flux of 0.5 mWb set up in it by a
     current of 10 A. Determine the self-inductance of the coil.
                                      [*Ans. 0.025 H*]

12. Calculate the energy stored in a magnetic circuit consisting of a coil of inductance 0.025 H when the current through the coil is 30 A.

[*Ans. 11.25 J*]

13. Calculate the energy stored in a magnetic circuit consisting of a coil of 5000 turns carrying a current of 5 A which is producing a magnetic flux of 1.6 mWb.

[*Ans. 20 J*]

14. A solenoid 20 cm long is uniformly wound with 5000 turns of wire on a form 5 cm in diameter. Calculate the inductance of the solenoid.

[*Ans. 0.308 H*]

15. A lifting magnet has a total core flux of 0.015 Wb and the area of each pole is 120 cm². Calculate the attractive force in newtons and kilograms it will exert on a load of magnetic material having sufficient section to occupy both pole faces.

[*Ans. 14918.85 N, 1520.78 kgf*]

16. The magnetic circuit of an electromagnet is 30 cm long and has a cross-sectional area of 3 cm²; the electromagnet has a coil of 500 turns. A current of 0.9 A was required to produce the necessary flux in the electromagnet. The B-H data for the iron are as follows:

H [AT/m]	500	800	1500
B [T]	0.8	1.0	1.2

Calculate the flux density in the magnetic circuit.

[*Ans. 1.2 T*]

17. Refer to Problem 16. To increase the rate of production of the electromagnets, the core was cut into two pieces. As a result, two gaps were introduced in the magnetic circuit, each of length 0.5 mm. Determine the increase in current necessary to maintain the flux density at the required level (1.2 T).

Neglect leakage and fringing fluxes.

[*Ans. 1.91 A*]

18. An iron ring wound with a coil of 120 turns of wire has a mean circumference of 50 cm and an air gap 1 mm long. Calculate the flux density when a current of 2.5 A flows through the coil. Assume a relative permeability of 300 for the iron.

[*Ans. 155.1 mT*]

19. A mild steel ring is uniformly wound with a coil of 200 turns. The ring has a cross-sectional area of 500 mm² and mean circumference of 400 mm. Calculate

    (a)   the reluctance of the ring

    (b)   the current required to establish a flux of 800 μWb in the ring. Assume a relative permeability of 380 for the material.

[Ans. *(a) 1.676 x 10⁶ AT/Wb, (b) 6.705 A*]

20. The magnetic circuit shown in Fig. 3.P-20 has a coil of 500 turns on the central limb, and the magnetic material is Stalloy. Calculate the current required in the coil to produce a flux density of 1.0 T in the air gap. Use the B/H curves in Fig. 3.11-1. (Hint: Answer may not exactly coincide as shown but should not differ by ± 10%.)   *[Ans. 1.6 A]*

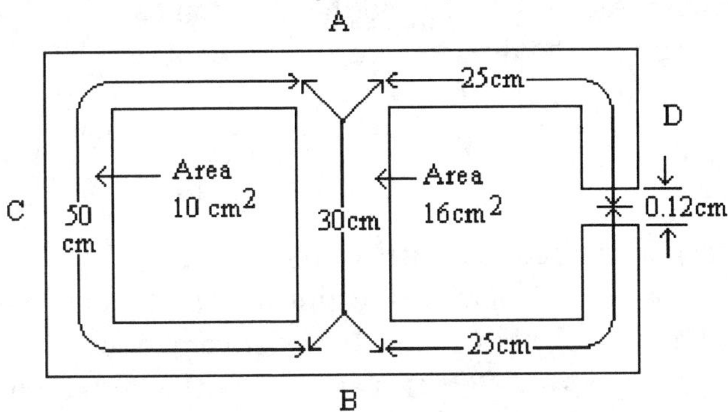

Fig. 3.P-20

[Hint: The total flux in the central limb is the sum of the fluxes in the two outer limbs.
(i) Calculate the m.m.f for the air gap; (ii) obtain flux density from B/H curve;
(iii) obtain m.m.f for right-outer limb. (iv) The m.m.f for path ADB = sum
of m.m.f of right-outer limb + m.m.f of gap; the m.m.f for outer limb ACB
is the same for path ADB.]

21. A coil of 500 turns is uniformly wound on a Stalloy ring of mean diameter 20 cm and cross-sectional area 3 cm². A radial gap of length 0.1 mm exists in the ring. Calculate (a) the direct current required to establish a flux of 405 μWb in the gap and (b) the relative permeability of the Stalloy at the working flux density. (Use the B/H curves in Fig. 3.11-1 and ignore magnetic leakage and fringing.)   *[Ans. (a) 1.47 A, (b) 1070]*

22. Calculate the approximate value of m.m.f required to produce a flux of 800 μWb in each air gap in the symmetrical magnetic circuit shown in Fig. 3.P-22. All dimensions are in centimeters and the iron is 5 cm thick throughout. The material is cast steel with data shown in Fig. 3.11-1 B/H curves.   *[Ans. 1620 AT]*

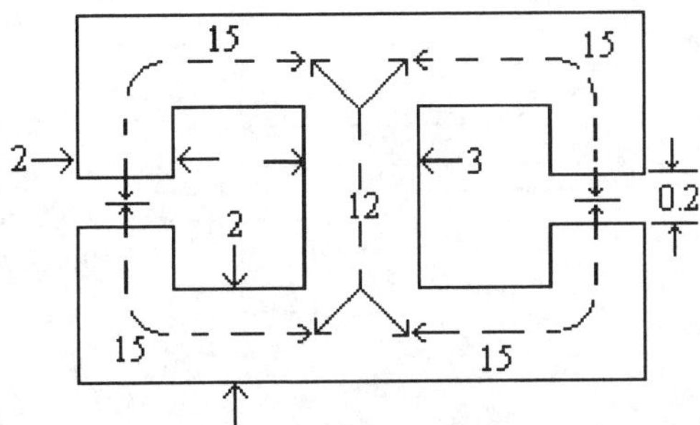

Fig. 3.P-22

23. A lifting magnet constructed of cast steel is wound with 400 turns. The magnet is shaped as an inverted "U" of mean length 2 m and cross-sectional area 25 cm². The magnet is to lift a cast iron bar of cross-sectional area 80 cm² and the mean length of the flux path through the load is 1.0 m. An air gap of length 0.5 mm and effective area 30 cm² exists between each pole face and the cast iron bar. The leakage coefficient is 1.1. Estimate (a) the direct current required to establish a flux density of 1.0 T in each air gap and (b) the total pull exerted by the magnet on the load in (i) newtons and (ii) pounds. Use Fig. 3.11-1 B/H curves.

[Ans. (a) 14.4 A, (b) 2390 N, 538 lb]

24. A magnetic circuit consists of three parts, A, B and C, in series and has in each part a uniform cross-sectional area of 4 cm². Part A has a magnetic path of mean length 15 cm and relative permeability of 1000 under the stated conditions. Part B is an air gap of length 2 mm. Part C has a mean magnetic path of 50 cm and magnetization characteristics as follows:

H (AT/m)	400	600	800	1000	1200	1400	1600	1800
B (T)	0.7	0.8	1.0	1.075	1.132	1.18	1.215	1.25

Determine the m.m.f required to set up a flux of 440 μWb in the circuit and the current to establish this m.m.f when flowing in a uniformly wound coil of 800 turns. Neglect magnetic leakage and fringing.

[Ans. 2428 AT, 3.035 A]

25. Given that the energy stored in a non-magnetic medium is $B^2/2\mu_o$ J/m³, derive an expression for the tractive force between two magnetized pole faces. The arm of a d.c motor starter is held in the "ON" position by means of an electromagnet (refer to Fig. 3.P-25). A spiral spring exerts an average countertorque of 8 N m on the armature in this position. The length between the center of the armature and the pivot on the starter arm is 20 cm and the cross-sectional area of each pole face of the electromagnet is 3.5 cm². Calculate the minimum m.m.f required on the electromagnet to keep the arm in the "ON" position when each air gap between the armature and the electromagnet is 0.5 mm. Assume the reluctance of the iron path is negligible and ignore the effects of magnetic leakage and fringing.

[Ans. 301 AT]

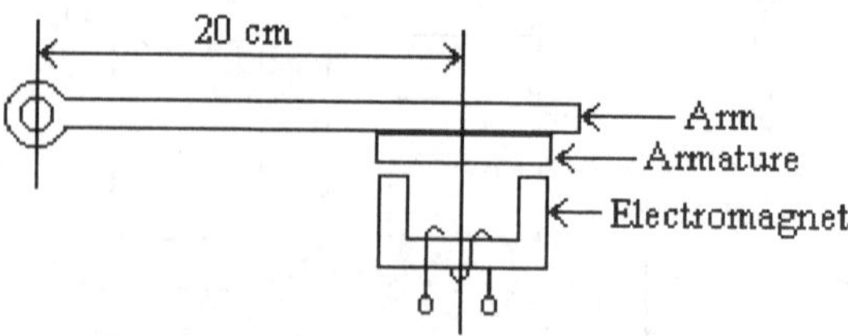

Fig. 3.P-25

26. A coil of 80 turns produces a flux of 0.1 mWb when carrying a current of 4 A. Calculate the inductance of the coil and the average e.m.f induced when this current is reversed in 0.01 sec. Comment on the direction of the induced e.m.f.

[*Ans. 2 mH, +1.6 V (refer to Example 3.24 for comments)*]

27. Two identical coils A and B, each uniformly wound with 1500 turns, lie in parallel planes such that 80% of the flux produced by one coil links with the other. A current of 4 A in coil A produces a flux of 60 μWb. If this current is reversed in 0.05 sec, determine the value of the voltage induced in coil B. Also calculate the self-inductance of each coil and the mutual inductance between them.

[*Ans. 2.88 V, 0.0225 H (of either coil), 0.018 H*]

28. A non-magnetic ring of mean diameter 20 cm and cross-sectional area 5 cm^2 is uniformly wound with 1000 turns. The current through the coil is 3 A. Calculate (a) the magnetizing force, (b) the flux density, (c) the flux, (d) the inductance, and (e) the average e.m.f induced in the circuit when the current is reversed in 0.05 sec.

[*Ans. (a) 4770 AT/m, (b) 0.006 T, (c) 3 μWb, (d) 1.0 mH, (e) +12 V*]

29. An air-cored toroid has a mean diameter of 30 cm and cross-sectional area of 3 cm^2. The toroid is uniformly wound with 450 turns. Calculate (a) the inductance of the coil and (b) the average e.m.f induced in the coil when a current of 1 A is reversed in 0.02 sec.

[*Ans. (a) 81 μH, (b) 8.1 mV*]

30. (a) What losses occur in a ferromagnetic material when it is subjected to alternating magnetization?
   (b) Why is it necessary to laminate a transformer core?
   (c) The core of a transformer has a volume of 0.16 m^3 and the iron loss was found to be 2170 W at 50 Hz. The hysteresis loop of the core material taken to the same maximum flux density had an area of 9.0 cm^2 when drawn to scales of 1 cm = 0.1 T and 1 cm = 250 AT/m. Calculate the total iron loss in the transformer core when it is energized to the same flux density but from a 60 Hz supply. [*Ans. 2693 W*]

31. A sample of iron has a hysteresis loop plotted to the following scales:
   Flux density,   1 in   = 0.1 T
   Magnetizing force,   1 in   = 50 AT/m
   The area of the complete loop is 31.46 in^2 for a maximum flux density. If the density of the sample iron is 7800 kg/m^3, determine the hysteresis loss in W/kg at this particular flux density. The frequency of cyclic magnetization is 50 Hz. [*Ans. 1.01 W/kg*]

32. A non-magnetic ring has a mean circumference of 60 cm and a mean diameter of 2.26 cm. It is uniformly wound with two coils, A and B. Coil A has 120 turns and coil B has 300 turns. Calculate (a) the mutual inductance of the coils and (b) the e.m.f induced in coil B when a current of 6 A in coil A is reversed in 0.01 sec. [*Ans. (a) 30 μH, (b) – 0.0362 V*]

33. Two coils have self-inductances of 400 μH and 250 μH, respectively. The coils are so positioned that their mutual inductance is 150 μH. Determine their combined self-inductance when the coils are coupled (a) cumulatively and (b) differentially. [Ans. *(a) 950 μH, (b) 350 μH*]

34. A coil has a resistance of 10 Ω and an inductance of 10 H. Determine the value of the current 0.1 sec after switching on a 100 V d.c. supply. Calculate also the time taken for the current to reach one-half of its steady value. [*Ans. 0.952 A, 0.697 sec*]

35. An electromagnetic relay has a coil of resistance 20 Ω and an inductance of 0.5 H. The relay is energized from a d.c voltage pulse of magnitude 15 V. The pulse rises from 0 to 15 V instantaneously, remains at 15 V for 0.30 sec, and then falls instantly to 0 V. The relay contacts close when the current attains 150 mA on the increase and open when the current is 80 mA decreasing. Determine the total time during which the contacts are closed. [*Ans. 0.2385 sec*]

36. A circuit breaker operates when the current through the trip coil has reached 40 % of its final steady value. The resistance and inductance of the trip coil are 250 Ω and 2 H, respectively, and it is energized from a 110 V d.c. source. Calculate (a) the final steady value of the current, (b) the time constant of the trip coil, (c) the initial rate of change of current, (d) the energy stored in the coil under steady-state condition, and (e) the time delay between energization and breaker operation.
[*Ans. (a) 0.44 A, (b) 8 ms, (c) 55 A/sec, (d) 0.194 J, (e) 4.09 ms*]

37. An air-cored coil of resistance 10 Ω and inductance 2 H is energized from a 50 V d.c source. Use an approximate method to construct a curve showing the rise of current with time. Hence, find the instantaneous value of current after time "t" equal to the time constant of the circuit. Compare the graphical value with the value obtained using the Helmholtz equation: $i = V/R\,[1 - e^{-Rt/L}]$ [*Ans. 3.16 A (by equation)*]

38. A coil of resistance 10 Ω and inductance 20 H is arranged such that it can be connected to either battery source A or battery source B by a change-over switch as shown in Fig. 3.P-38. Determine the current flowing and its direction (a) 0.4 sec, (b) 0.8 sec, and (c) 3 sec after changing over the switch from A to B. The terminal voltage of battery A is 200 V and that of B is 100 V. [*Ans. (a) +14.56 A, (b) +10.11 A, (c) –3.31 A*]

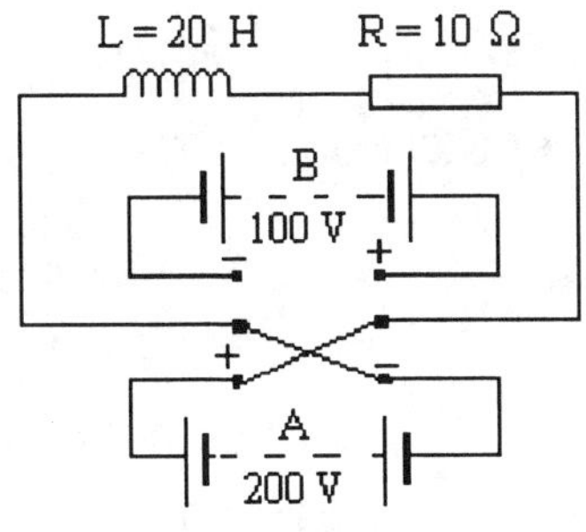

Fig. 3.P-38

39.   (a) A solenoid 1.2 cm long is uniformly wound with 720 turns. A search coil 5 cm in diameter and wound with 30 turns is placed centrally inside the solenoid; both coils are positioned in a vacuum. The search coil circuit has a resistance of 5000 Ω and is connected to a ballistic galvanometer having a constant of 2×10^{-9} coulombs per division. Calculate (i) the galvanometer deflection when a current of 5 A in the solenoid is reversed and (ii) the mutual inductance between the coils.

     (b) The search coil is now connected instead to a fluxmeter having a a constant of 15 μWb. Calculate the fluxmeter deflection when the current of 5 A in the solenoid is reversed.

[Ans. (a) (i) 44.4 divisions, (ii) 44.4 µH, (b) 26.9 divisions]

CHAPTER   4

ELECTROSTATICS

Introduction

Electrostatics : Nomenclature, Symbol and Units

Term	Symbol	Unit	
Quantity of charge	Q	Coulombs	[C]
Electric flux	ψ	Coulombs	[C]
Electric flux density	D	Coulombs/m^2	[C/m^2]
Applied voltage	V	Volts	[V]
Area of plates (electrodes)	A	Square meters	[m^2]
Distance plates are separated by thickness of dielectric	d	Meters	[m]
Electric field strength (or potential gradient or electric stress or electric field intensity)	E	Volts/meter	[V/m]
Permittivity of free space	ε_o	Farads/meter	[F/m]
Relative permittivity	ε_r	Farads/meter	[F/m]
Capacitance	C	Farads	[F]

The Capacitor

A capacitor basically consists of two metal plates separated by an insulating material. The two metal plates are better termed *electrodes* and the insulation *dielectric*.

When an electrical potential is applied across the electrodes, an *electric field* is established in the dielectric and the capacitor becomes charged. The capacitor will remain in this charged state until physically discharged.

The quantity of charge Q is proportional to the applied voltage. Thus

$$Q \propto V$$

and
$$Q = C V \text{ coulombs} \qquad \text{(Eq. 4.0-0)}$$

where Q = charge in coulombs

V = applied voltage in Volts

C = capacitance of capacitor in farads

The unit of capacitance C is the farad (symbol F) and is defined as follows:

The farad is the capacitance of a capacitor which stores a charge of one coulomb when a potential difference of one volt appears between the electrodes.

Capacitance of a parallel plate capacitor

The simplest form of capacitor is the parallel plate capacitor.
Consider a pair of parallel plates separated by free space where a p.d of V volts is applied across the plates as shown in Fig. 4.0-0

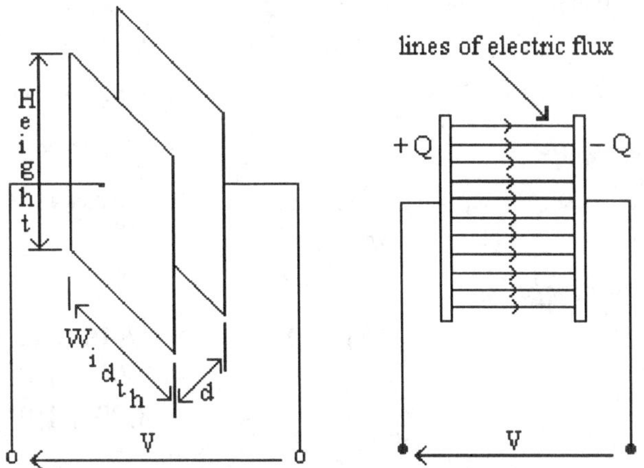

Fig. 4.0-0. A parallel plate capacitor

Area, A = width x height (m²)   d = distance between plates (m)

Dielectric material = free space or air   V = applied p.d

Refer to Fig. 4.0-0

The electric field strength, symbol E, is expressed as

$$E = {}^{V}/d \text{ volts/meter} \qquad \text{(Eq. 4.0-1)}$$

Assume that the lines of electric flux emanating from the positive (+) charge of Q coulombs enter the negative (–) charge of Q coulombs; then the electric flux between the plates is expressed thus:

$$\psi = Q \text{ coulombs}$$

The flux passes through a dielectric of area A. Hence, the flux density,

$$D = {}^{\psi}/A$$
$$= {}^{Q}/A \text{ coulombs/m}^2 \qquad \text{(Eq.4.0-2)}$$

Combining Eq. 4.0-1 and Eq. 4.0-2, we get

$$\frac{\underline{\text{Electric flux density}}}{\text{Electric field strength}} = {}^{D}/E$$

$$= {}^{Q}/A \div {}^{V}/d = {}^{Q}/A \times {}^{d}/V$$

$$= {}^{Qd}/AV \qquad = {}^{Cd}/A \quad (C = {}^{Q}/V)$$

(contd)

Cd/A is termed the permittivity of free space or electric space constant, symbol ε_o; that is

$$\varepsilon_o = Cd/A \qquad \text{(Eq. 4.0-3)}$$

The value of ε_o is 8.85×10^{-12} F/m.

There is a definite relationship between μ_o, ε_o and the velocity of light. Thus

$$\text{Permeability of vacuum } (\mu_o) = 4\pi \times 10^{-7} \text{ H/m}$$

$$\text{Velocity of light in vacuum } (c_o) = 3 \times 10^8 \text{ m/s}$$

$$\begin{aligned}
\text{Permittivity of vacuum } (\varepsilon_o) &= 1/(\mu_o c_o^2) \\
&= 1/[4\pi \times 10^{-7} (3 \times 10^8)^2] \\
&= 1/(4\pi \times 10^{-7} \times 9 \times 10^{16}) \\
&= 1/(36\pi \times 10^9)
\end{aligned}$$

i.e., $\qquad \varepsilon_o = \dfrac{1}{36\pi \times 10^9} = 8.85 \times 10^{-12}$ F/m

Alternatively

$$\begin{aligned}
1/(\mu_o \varepsilon_o) &= 1/(4\pi \times 10^{-7} \times 8.85 \times 10^{-12}) \\
&= 8.99 \times 10^{-3} \times 10^{19} \\
&= 8.99 \times 10^{16} = (2.998 \times 10^8)^2 \\
&\approx (3 \times 10^8)^2
\end{aligned}$$

But the velocity of light $(c_o) = 3 \times 10^8$ m/s

$\therefore \qquad$ velocity of light $(c_o) = 1/\sqrt{(\mu_o \varepsilon_o)}$.

From Eq. 4.0-3 $\qquad C = \varepsilon_o A/d \qquad$ farads (Eq. 4.0-4)

If the space between the capacitor plates is filled with a dielectric, then the expression for C in Eq. 4.0-4 is modified to

$$C = \varepsilon_o \varepsilon_r A/d \qquad \text{farads (Eq. 4.0-5)}$$

where ε_r = relative permittivity of the dielectric material.

When the plates are separated by a dielectric material, the value of the capacitance is increased by a factor equivalent to the relative permittivity of the dielectric material.

Capacitance of multiplate capacitor

Consider a capacitor made up of n parallel plates, alternate plates being connected together as shown in Fig. 4.0-1.

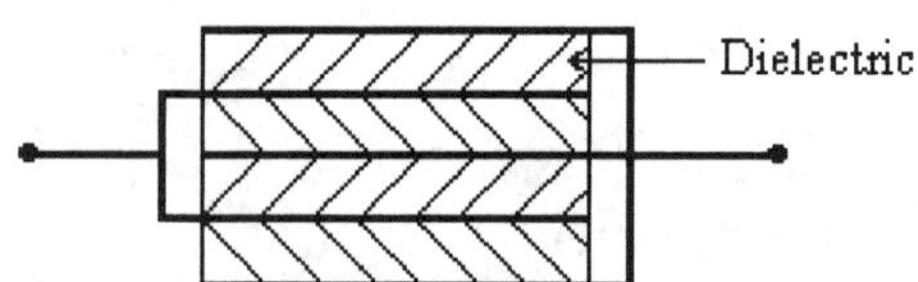

Fig. 4.0-1. Multiplate capacitor

<u>Refer to Fig. 4.0-1</u>

The multiplate capacitor consists of 5 plates separated by 4 dielectrics (the reason for 4 dielectrics being that only one side of each outer plate forms part of the capacitor). In the general case of an n-plate capacitor, there are $(n - 1)$ dielectrics and the effective area of a set of plates is $(n - 1)A$, where A is the area of one plate.

For a capacitor consisting of n plates, Eq. 4.0-5 is now modified to read

$$C = \frac{\varepsilon_o \varepsilon_r (n - 1) A}{d} \quad \text{farads} \quad \text{(Eq. 4.0-6)}$$

where ε_o = 8.85 x 10^{-12}

ε_r = relative permittivity of dielectric material

A = area of electrodes (plates) in m^2

d = distance plates are separated by dielectric thickness

Equation 4.0-6 has some very interesting information and I do not believe it would be inappropriate to make the introduction at this time, which is as follows:

"The capacitance C is directly proportional to the area (A) of the plates and inversely proportional to the distance (d) between the plates".
Therefore

$$(1) \quad C \quad \propto \quad A$$

Hence, increasing the area of the plates will increase the capacitance of a capacitor.

$$(2) \quad C \quad \propto \quad {}^1\!/d$$

That is, increasing the distance between the plates will decrease the capacitance of the capacitor. Consequently

To increase the value of capacitance in a circuit, capacitors are connected in parallel. This has the effect of increasing the area of the plates. When the capacitors are connected in series, the distance between the plates is increased, and as a result the capacitance will decrease.

Capacitors in parallel

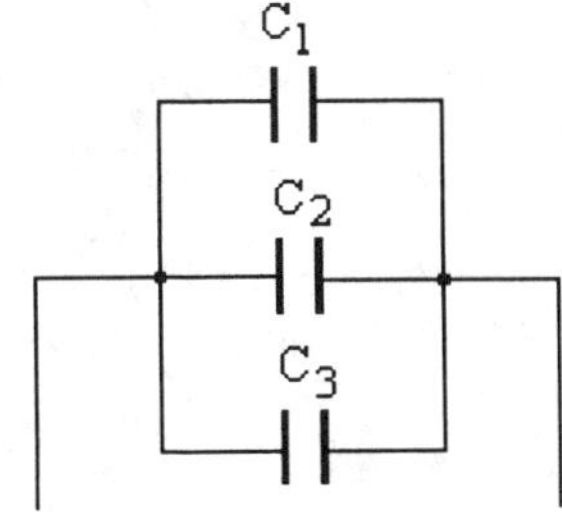

Fig. 4.0-2. Capacitors in parallel

<u>Refer to Fig. 4.0-2</u>

Total equivalent capacitance is

$$C_T = C_1 + C_2 + C_3 + \ldots C_n \qquad \text{(Eq. 4.0-7)}$$

Capacitors in series

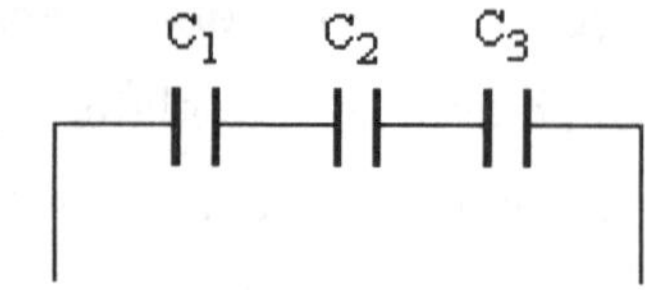

Fig. 4.0-3. Capacitors in series

<u>Refer to Fig. 4.0-3</u>

Total equivalent capacitance is

$$^1/C_T = {}^1/C_1 + {}^1/C_2 + {}^1/C_3 + \ldots {}^1/C_n \qquad \text{(Eq. 4.0-8)}$$

Voltage distribution of capacitors in series

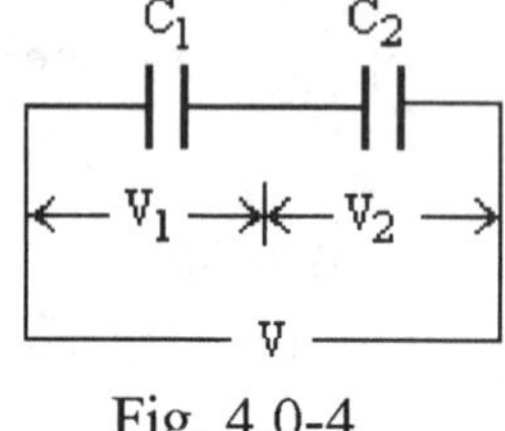

Fig. 4.0-4

<u>Refer to Fig. 4.0-4</u>

Since the capacitors are in series, the charge on each capacitor is equal to the total charge. Therefore

$$Q = C_1 V_1 = C_2 V_2$$

and $\quad V_1 = {}^Q/C_1 : \quad V_2 = {}^Q/C_2$

from which $\quad V_1/V_2 = {}^{C_2}/C_1 \qquad \text{(Eq. 4.0-9)}$

But $\quad V = V_1 + V_2$

$\therefore \quad V_1 = V - V_2$

Substituting for V_1 in Eq. 4.0-9, we get

$$(V - V_2)/V_2 = {}^{C_2}/C_1$$
$$VC_1 - V_2 C_1 = V_2 C_2$$
$$V_2 C_1 + V_2 C_2 = VC_1$$
$$V_2(C_1 + C_2) = VC_1$$
$$V_2 = V \times \frac{C_1}{C_1 + C_2} \qquad \text{(Eq. 4.0-10)}$$

Similarly, $\quad V_1 = V \times \dfrac{C_2}{C_1 + C_2} \qquad \text{(Eq. 4.0-11)}$

Energy stored in a capacitor

The fundamental property of a capacitor is to store electrical energy. It does not dissipate energy; that is, no energy is lost.

The energy stored in a capacitor is given by the following expression:

$$\text{Energy stored, } W = \tfrac{1}{2} CV^2 \text{ Joules} \qquad \text{(Eq. 4.0-12)}$$

Types of capacitors

Basically there are two types of capacitors, which may be subdivided into groups according to the physics of the dielectric. The two types are

(1)   Polarized capacitors– Appropriate in circuits where the supply to the capacitor never reverses polarity

(2)   Nonpolarized capacitors– Mainly used in high-frequency circuits

Polarized capacitors are generally called electrolytic capacitors and are commonly used in d.c circuits, for example, as a filter in a rectifier circuit to reduce or filter the ripple in the rectified wave.

Charging and discharging of a capacitor

When a capacitor is charged through a resistor of R ohms from a p.d of V volts, the instantaneous voltage v across the capacitor C is given by

$$v = V(1 - e^{-t/CR}) \text{ volts} \qquad \text{(Eq. 4.0-13)}$$

The instantaneous charging current is given by

$$i = (V/R)e^{-t/CR}$$

where $V/R = I$

$$\therefore \quad i = Ie^{-t/CR} \text{ ampere} \qquad \text{(Eq. 4.0-14)}$$

When a capacitor of C farads charged to a p.d of V volts is discharged through a resistor of R ohms, the instantaneous discharge voltage and current are given by

$$v = Ve^{-t/CR} \text{ volts} \qquad \text{(Eq. 4.0-15)}$$

$$\text{and} \quad i = Ie^{-t/CR} \text{ ampere} \qquad \text{(Eq. 4.0-16)}$$

where e is an exponential constant and has a value of 2.718

and CR = time constant of the circuit in seconds.

Time constant

The time constant CR of a circuit may be defined as the time required for the voltage across the capacitor to raise from 0 to 63.2 % of its final value. _Alternatively_, the time constant (CR seconds) may be defined as the time required for the voltage across the capacitor C to increase from 0 to its final value if it was to continue to increase at its initial rate.

Summary

1. Electric flux

$$\psi = Q \quad \text{coulombs} \quad [C]$$

2. Electric flux density

$$D = \psi/A = Q/A \quad [C/m^2]$$

3. Electric field strength (also described as potential gradient or electric field intensity or electric stress)

$$E = V/d \quad [V/m]$$

4. Capacitance of parallel plate capacitor

$$C = \varepsilon_o \varepsilon_r \, A/d \quad [F]$$
$$\varepsilon_o = 1/(36\pi \times 10^9)$$
$$= 8.85 \times 10^{-12} \quad [F/m]$$
$$\varepsilon_r = \text{value depends on nature of dielectric}$$

5. Parallel plate capacitor with n plates

$$\text{Capacitance, } C = \frac{\varepsilon_o \varepsilon_r \, (n-1) \, A}{d} \quad [F]$$

6. Capacitors in parallel

$$C_T = C_1 + C_2 + C_3 + \dots C_n$$

7. Capacitors in series

$$1/C_T = 1/C_1 + 1/C_2 + 1/C_3 + \dots 1/C_n$$

When two capacitors C_1 and C_2 are connected in series

$$C_T = \frac{C_1 C_2}{C_1 + C_2}$$

8. Energy stored in a capacitor

$$W = \tfrac{1}{2} CV^2 \quad [J]$$

9. Instantaneous voltage or current in a capacitive circuit when

Charging $\quad v = V(1 - e^{-t/CR}) \quad [V]$

$\qquad\qquad\quad i = Ie^{-t/CR} \quad [A]$

Discharging $\quad v = Ve^{-t/CR} \quad [V]$

$\qquad\qquad\qquad i = Ie^{-t/CR} \quad [A]$

Example 4.1

*A voltage of 400 V d.c. is maintained across the electrodes of a capacitor
having a capacitance of 200 μF. Calculate (a) the charge stored and
(b) the potential gradient in the dielectric given that its thickness is 2 mm.*

Solution

(a)

$$\text{Charge, } Q = CV \quad \text{coulombs}$$
$$1 \ \mu F = 10^{-6} \ F$$
$$\therefore \quad Q = 200 \times 10^{-6} \times 400$$
$$= 8 \times 10^{-2} \qquad = \underline{\mathbf{0.08 \ C}}$$

(b)   The potential gradient or electric field strength is

$$E = {}^{V}/d$$
$$= {}^{400}/(2 \times 10^{-3})$$
$$= 200 \times 10^{3} \qquad = \underline{\mathbf{200 \ kV/m}}$$

Example 4.2

*A parallel plate capacitor consists of two metal plates, each of
dimensions 10 cm x 50 cm and separated by a slab of dielectric material
1.5 mm in thickness and having a relative permittivity of 5.0. Determine
(a) the capacitance of the capacitor, (b) the charge stored, (c) the electric
field strength and (d) the electric flux density if the voltage between the
plates is 400 V.*

Solution

(a)

$$\text{Capacitance, } C = \varepsilon_o \varepsilon_r \ {}^{A}/d \quad \text{[refer to Eq. 4.0-5]}$$
$$\text{Area of plate, } A = 10 \text{ cm x } 50 \text{ cm} \quad = 500 \times 10^{-4} \ m^2$$
$$\varepsilon_o = 8.85 \times 10^{-12}$$
$$\varepsilon_r = 5.0$$
$$d = 1.5 \text{ mm} = 1.5 \times 10^{-3} \ m$$
$$\therefore \quad C = \frac{8.85 \times 10^{-12} \times 5.0 \times 500 \times 10^{-4}}{1.5 \times 10^{-3}}$$
$$= \frac{8.85 \times 5.0 \times 500 \ \times 10^{-13}}{1.5}$$
$$= 14750 \times 10^{-13} \times 10^{12} \ pF$$
$$= 14750 \times 10^{-1} \qquad = \underline{\mathbf{1475 \ pF}}$$

(contd)

(b) Charge stored, Q $= CV$ [refer to Eq.4.0-0]

$$= 1475 \times 10^{-12} \times 400 \ C$$

$$= 1.475 \times 0.4 \times 10^{-6}$$

$$= 1.475 \times 0.4 \qquad = \underline{\mathbf{0.59 \ \mu C}}$$

(c) Electric field intensity

$$E \quad = {}^{V}/_{d}$$

$$= {}^{400}/_{(1.5 \times 10^{-3})} \qquad = \underline{\mathbf{266.67 \ kV/m}}$$

(d) Electric flux density

$$D \quad = {}^{Q}/_{A} \qquad \text{[refer to Eq. 4.0-2]}$$

$$= \frac{0.59 \times 10^{-6}}{500 \times 10^{-4}}$$

$$= 1.18 \times 10^{-5} \ C/m^2$$

$$= 1.18 \times 10^{-5} \times 10^{6} \quad \mu C/m^2$$

$$= \underline{\mathbf{11.8 \ \mu C/m^2}}$$

Example 4.3

A capacitor is constructed of 5 metal plates connected as shown in Fig. 4.0-1 and separated by mica sheets of thickness 0.5 mm and relative permittivity 3.5. The area of one side of each plate is 500 cm². Calculate the capacitance in microfarads.

Solution

Refer to Eq. 4.0-6

$$C \quad = \frac{\varepsilon_0 \varepsilon_r \ (n-1) \ A}{d}$$

where $\varepsilon_0 \ = 8.85 \times 10^{-12}$

$\varepsilon_r \ = 3.5$

$(n - 1) \ = 5 - 1 = 4$

$A \ = 500 \times 10^{-4}$

$d \ = 0.5 \times 10^{-3}$

$$\therefore \quad C \quad = \frac{8.85 \times 10^{-12} \times 3.5 \times 4 \times 500 \times 10^{-4}}{0.5 \times 10^{-3}}$$

$$= 123900 \times 10^{-13} \ F$$

$$= 12390 \times 10^{-12} \ F$$

$$= 0.012390 \times 10^{-6} \ \mu F \qquad \approx \underline{\mathbf{0.0124 \ \mu F}}$$

Example 4.4

A capacitor made up of a composite dielectric has two 16 cm^2 metal plates spaced 8 mm apart. The dielectrics separating the plates are mica and Bakelite of thickness 2 mm and 6 mm, respectively. The relative permittivity of the mica is 4.0 and that of the Bakelite is 5.0. Calculate (a) the capacitance and (b) the electric field strength of each dielectric when a p.d of 15 kV is applied across the metal plates.

Solution

(a) Consider Fig. 4.4-0, which shows the structure of the capacitor (diagram not to scale).

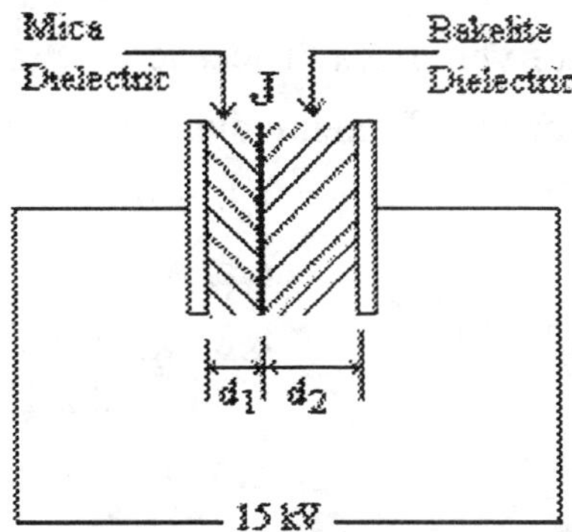

Fig. 4.4-0. Parallel plate capacitor
with composite dielectric

<u>Refer to Fig. 4.4-0</u>

Let   J   = junction surface where the two dielectrics meet

d_1 = thickness of the mica dielectric   = 2 mm

d_2 = thickness of the Bakelite dielectric = 6 mm

A   = area of each dielectric = 16 cm^2 = 16 x 10^{-4} m^2

E_1 = electric field strength of mica dielectric

E_2 = electric field strength of Bakelite dielectric

The distance between the plates is

$$= d_1 + d_2$$
$$= 2 + 6 = 8 \text{mm} = 8 \text{ x } 10^{-3} \text{ m}$$

The potential gradient between the plates is

$$= 15 \text{ kV}$$

The p.d between a dielectric (mica or Bakelite) is

$$= \text{electric field strength x thickness}$$
$$= E_1 d_1 \text{ or } E_2 d_2$$

(contd)

Hence any point on the junction surface J is at the same potential; that is, the surface J is an equipotential surface. Consequently, the construction may be regarded as two capacitors C_1 and C_2 connected in series as shown in Fig. 4.4-1.

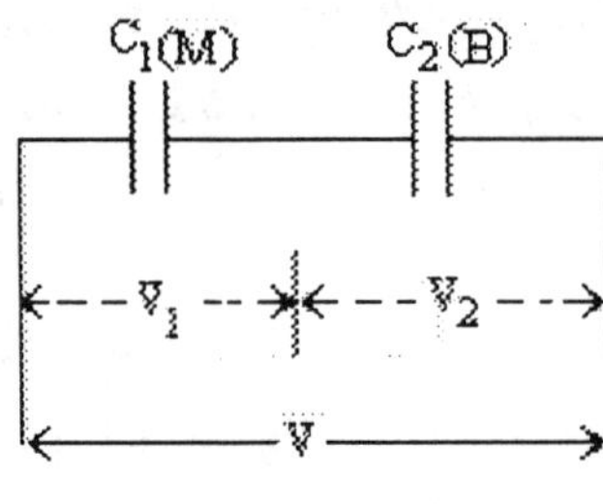

Fig. 4.4-1

$C_{1(M)}$ = capacitor with mica dielectric   ($\varepsilon_{r(M)}$ = 4)

$C_{2(B)}$ = capacitor with Bakelite dielectric   ($\varepsilon_{r(B)}$ = 5)

$$C_1 = \varepsilon_o\varepsilon_r\, A/d_1$$

$$C_2 = \varepsilon_o\varepsilon_r\, A/d_2$$

$$\therefore \quad C_1 = \frac{8.85 \times 10^{-12} \times 4 \times 16 \times 10^{-4}}{2 \times 10^{-3}}$$

$$= \frac{8.85 \times 64 \times 10^{-13}}{2} \qquad = \underline{28.32 \text{ pF}}$$

$$C_2 = \frac{8.85 \times 10^{-12} \times 5 \times 16 \times 10^{-4}}{6 \times 10^{-3}}$$

$$= \frac{8.85 \times 80 \times 10^{-13}}{6} \qquad = \underline{11.8 \text{ pF}}$$

The equivalent capacitance is given by

$$C_T = \frac{C_1 C_2}{C_1 + C_2} \qquad \text{[refer to summary no. 7]}$$

$$= \frac{28.32 \times 11.8}{28.32 + 11.8} \qquad = \underline{\mathbf{8.33 \text{ pF}}}$$

(b) Since the capacitors are in series, Eq. 4.0-10 or Eq. 4.0-11 may be applied. Therefore, the potential gradients across C_1 and C_2 are

$$V_1 = V \times \frac{C_2}{C_1 + C_2}$$

$$= 15 \times 10^3 \times \frac{11.8 \times 10^{-12}}{(28.32 + 11.8) \times 10^{-12}}$$

$$= \frac{15 \times 10^3 \times 11.8}{40.12} \qquad = \underline{\mathbf{4.41 \text{ kV}}}$$

$$V_2 = V - V_1$$

$$= (15 - 4.41) \times 10^3 \qquad = \underline{\mathbf{10.59 \text{ kV}}}$$

Example 4.5

A capacitor is composed of two flat parallel plates separated by a dielectric of thickness 0.5 mm. The plates of each capacitor are rectangular, measuring 15 cm x 10 cm and having a capacitance of 500 pF. Calculate the dielectric flux density and electric stress when the capacitor is connected to 100 Vd.c. What would be the capacitance if the thickness of the dielectric is increased to 1.0 mm?

Solution

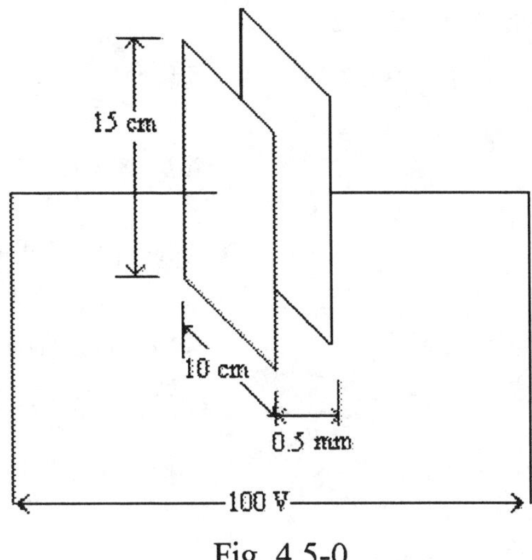

Fig. 4.5-0

<u>Refer to Fig. 4.5-0</u>

Electric flux density, D $= \Psi/A$

$$= Q/A = CV/A$$

$$= \frac{500 \times 10^{-12} \times 100}{15 \times 10 \times 10^{-4}}$$

$$= \frac{500}{150} \times 10^{-6} \qquad = \underline{\mathbf{3.33\ \mu C/m^2}}$$

Electric stress, E $= V/d = \dfrac{100}{0.5 \times 10^{-3}} \qquad = \underline{\mathbf{200\ kV/m}}$
$\underline{\mathbf{or\ 200\ V/mm}}$

Dielectric thickness is increased to 1.0 mm; that is, the thickness is doubled.

<u>Refer to Eq. 4.0-4</u>

$$C \quad = \varepsilon_o A/d$$

From the interpretation of this equation, doubling the dielectric thickness will reduce the capacitance by half; hence

$$C \quad = 500/2 \qquad = \underline{\mathbf{250\ pF}}$$

Example 4.6

A multiplate capacitor has 10 metal plates, each 15 cm x 15 cm. The plates are separated by mica having a thickness of 0.3 mm and relative permittivity 5.6. Calculate

(a)   the capacitance of the capacitor so formed

(b)   the charge on the capacitor in microcoulombs when connected to a 400 Vd.c. supply

Solution

(a)   Using Eq. 4.0-6, we get

$$\text{Capacitance, } C = \frac{\varepsilon_o \varepsilon_r \, (n-1) \, A}{d}$$

$$= \frac{8.85 \times 10^{-12} \times 5.6 \, (10-1) \times 15^2 \times 10^{-4}}{0.3 \times 10^{-3}}$$

$$= \frac{8.85 \times 5.6 \times 9 \times 15^2 \times 10^{-13}}{0.3} \times 10^6 \, \mu F$$

$$= 334530 \times 10^{-7} \qquad = \mathbf{0.03345 \, \mu F}$$

(b)   Charge, $Q = C V = 0.03345 \times 400 \qquad = \mathbf{13.38 \, \mu F}$

Example 4.7

Two capacitors, 10 μF and 25 μF, are connected in parallel and the combination is fed from a 250 V d.c supply. Calculate (a) the total capacitance, (b) the total charge and (c) the charge on each capacitor.

Solution

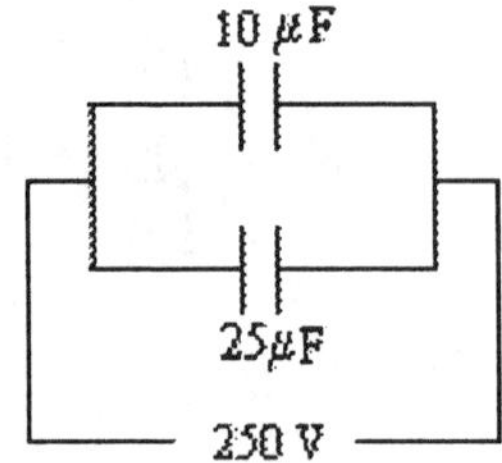

Fig. 4.7-0. Capacitors in parallel

Refer to Fig. 4.7-0

(a)   When capacitors are connected in parallel, the total capacitance is given by Eq. 4.0-7.

$$C_T = C_1 + C_2$$

Let $C_1 = 10 \, \mu F$ and $C_2 = 25 \, \mu F$

$$\therefore \qquad C_T = 10 + 25 \qquad = \mathbf{35 \, \mu F}$$

(contd)

(b)   Total charge   Q $= C\,V$

$$= 35 \times 10^{-6} \times 250 \qquad = \underline{\mathbf{8750\ \mu C}}$$

$$\underline{\mathbf{or\ 0.00875\ C}}$$

(c)   Charge Q on each capacitor is as follows:

On 10 μF, charge Q $= C\,V$

$$= 10 \times 10^{-6} \times 250 \qquad = \underline{\mathbf{0.0025\ C}}$$

On 25 μF, charge Q $= 25 \times 10^{-6} \times 250 \qquad = \underline{\mathbf{0.00625\ C}}$

Example 4.8

Two capacitors, 20 μF and 30 μF, are connected in series to a 1000 Vd.c. supply. Calculate the total capacitance of the combination and the electrical charge stored in the combination.

Solution

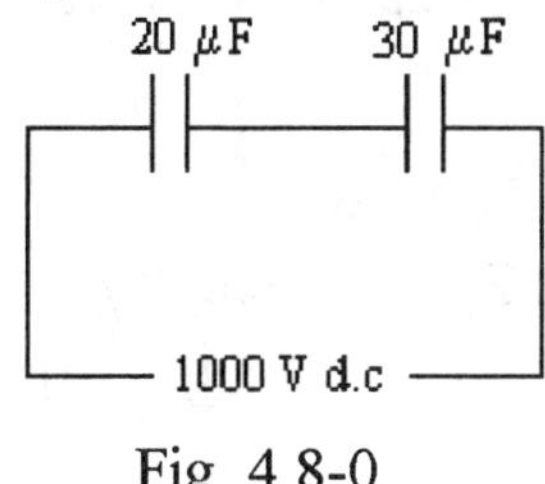

Fig. 4.8-0

<u>Refer to Fig. 4.8-0</u>

Let   20 μF $= C_1$

       30 μF $= C_2$

Total capacitance, $C_T \quad = \dfrac{C_1\,C_2}{C_1 + C_2}$

$$= \frac{20 \times 30}{20 + 30}\ \mu F \qquad = \underline{\mathbf{12\ \mu F}}$$

<u>*Alternatively*</u>

$$\frac{1}{C_T} = \frac{1}{C_1} + \frac{1}{C_2} = \frac{C_1 + C_2}{C_1\,C_2}$$

Inverting $C_T \quad = \dfrac{C_1\,C_2}{C_1 + C_2}$

Electrical charge stored in the combination is

$$Q \quad = C_T V$$

$$= 12 \times 10^{-6} \times 1000 \qquad = \underline{\mathbf{12000\ \mu C}}$$

$$\underline{\mathbf{or\ 0.012\ C}}$$

Example 4.9

Two parallel plate capacitors have the following dimensions:

Capacitor C_1:	*Area of plates*	*1600 cm²*
	Dielectric	*mica*
	Thickness of dielectric	*3 mm*
	Relative permittivity	*6*
Capacitor C_2:	*Area of plates*	*2500 cm²*
	Dielectric	*Glass*
	Thickness of dielectric	*5 mm*
	Relative permittivity	*3.5*

Calculate (a) the capacitance of each capacitor, (b) the total capacitance when C_1 and C_2 are connected in series, (c) the voltage across each capacitor when they are connected in series, and (d) the potential gradient across each of the dielectrics

A 500 V d.c. supply is applied to the series combination.

Solution

(a)

$$\text{Capacitance of } C_1 = \varepsilon_o \varepsilon_r \, A/d$$

$$= \frac{8.85 \times 10^{-12} \times 6 \times 1600 \times 10^{-4}}{3 \times 10^{-3}}$$

$$= 8.85 \times 2 \times 160 \times 10^{-12} \qquad = \underline{\mathbf{2832 \ pF}}$$

$$\text{Capacitance of } C_2 = \frac{8.85 \times 10^{-12} \times 3.5 \times 2500 \times 10^{-4}}{5 \times 10^{-3}}$$

$$= 8.85 \times 3.5 \times 50 \times 10^{-12} \qquad = \underline{\mathbf{1548.75 \ pF}}$$

(b)

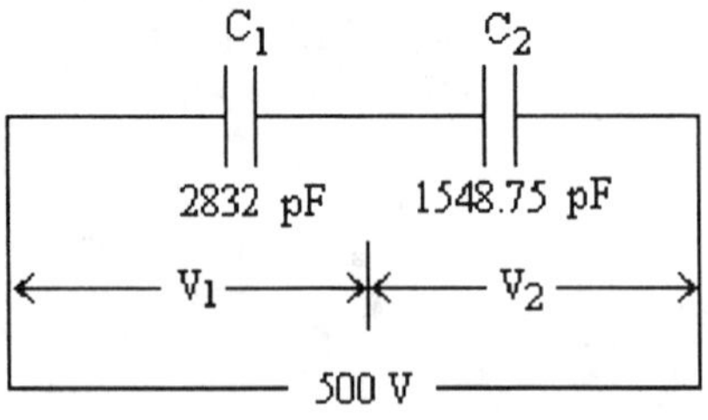

Fig. 4.9-0

<u>Refer to Fig. 4.9-0</u>

$$\text{Total capacitance, } C_T = \frac{C_1 \, C_2}{C_1 + C_2}$$

(contd)

$$= \frac{(2832 \times 1548.75) \times 10^{-24}}{(2832 + 1548.75) \times 10^{-12}}$$

$$= \frac{4386060 \times 10^{-12}}{4380.75} \quad = \mathbf{10001.2 \ pF}$$

(c) When the two capacitors are connected in series, their individual p.ds are in inverse proportion to their capacitances. Hence

$$\text{p.d across } C_1 \quad = V_1 \quad = V \times \frac{C_2}{C_1 + C_2} \quad \text{(refer to Eq. 4.0-11)}$$

$$= 500 \times \frac{1548.75}{4380.75} \quad = \mathbf{176.77 \ V}$$

$$\text{p.d across } C_2 \quad = V_2 \quad = V \times \frac{C_1}{C_1 + C_2} \quad \text{(refer to Eq. 4.0-10)}$$

$$= 500 \times \frac{2832}{4380.75} \quad = \mathbf{323.23 \ V}$$

(d) Potential gradient across $C_1 \quad = V/d \ = 176.77/(3 \times 10^{-3}) \ = \mathbf{58.92 \ kV/m}$

Potential gradient across $C_2 \quad = V/d \ = 323.23/(5 \times 10^{-3}) \ = \mathbf{64.65 \ kV/m}$

Example 4.10

A bank of three capacitors connected in series has a total capacitance of 2 μF. The known values of two of the capacitors are 4 μF and 8 μF. Calculate the value of the third capacitor and the voltage across each when 240 V d.c. is applied across the series combination.

Solution

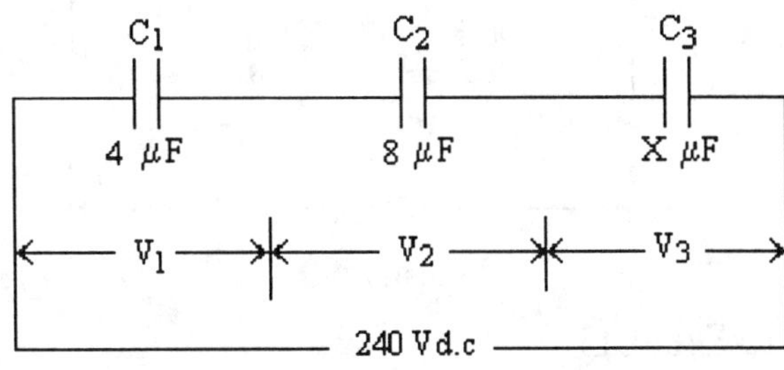

Fig. 4.10-0

<u>Refer to Fig. 4.10-0</u>

$$1/C_T \ = \ 1/C_1 + 1/C_2 + 1/C_3$$
$$1/C_3 = \ 1/C_T - 1/C_1 - 1/C_2$$
$$= \ 1/2 \ - 1/4 \ - 1/8 \ = \ 1/8$$

(contd)

$$\therefore \text{(Inverting)} \quad C_3 = (\text{invert } ^1/8) = \underline{\mathbf{8\ \mu F}}$$

$$\text{Total charge,} \quad Q = C_T V$$
$$= 2 \times 240 = 480 \times 10^{-6}\ C$$

Since the capacitors are in series, the charge in each is equal to the total charge, Q.

$$\therefore \quad V_1 = Q/C_1 = {}^{480}/_4 = \underline{\mathbf{120\ V}}$$
$$V_2 = Q/C_2 = {}^{480}/_8 = \underline{\mathbf{60\ V}}$$
$$V_3 = Q/C_3 = {}^{480}/_8 = \underline{\mathbf{60\ V}}$$

Example 4.11

Two parallel plate capacitors C_1 and C_2 are connected in series to a 240 V d.c. supply. The capacitance of C_1 is 0.02 μF and when C_1 is fully charged, the potential gradient across it is 100 V. The distance between the plates (of C_1) is 1.5 mm. The construction of capacitor C_2 is identical to that of C_1 except for the distance between the plates.

Calculate

> *(i)   the capacitance of C_2*
>
> *(ii)   the total capacitance*
>
> *(iii)   the distance between the plates of C_2*
>
> *(iv)   the charge on each capacitor.*

Solution

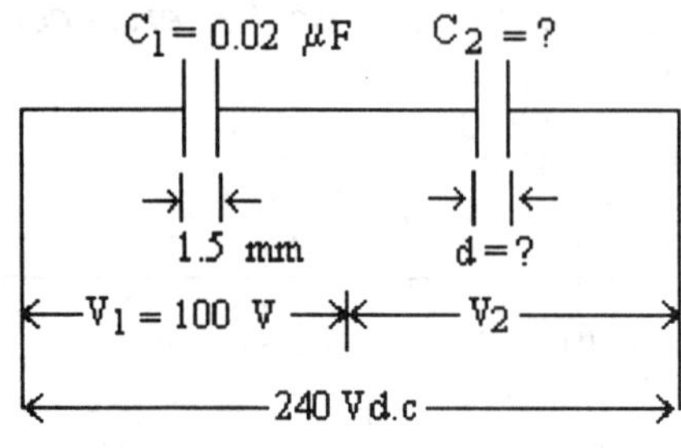

Fig. 4.11-0

Refer to Fig. 4.11-0

(i)

$$C_1/C_2 = V_2/V_1$$
$$\therefore \quad C_2 = C_1 V_1/V_2$$
$$V_2 = V\text{d.c} - V_1$$
$$= 240 - 100 = \underline{140\ V}$$
$$\therefore \quad C_2 = \frac{0.02 \times 100}{140} = \underline{\mathbf{0.0143\ \mu F}}$$

(contd)

(ii) Total capacitance

$$C_T = \frac{C_1 C_2}{C_1 + C_2}$$

$$= \frac{0.02 \times 0.0143}{0.02 + 0.0143} \qquad = \underline{\mathbf{0.0083\ \mu F}}$$

(iii) $C_1 = \varepsilon_o A/d_1 \qquad C_2 = \varepsilon_o A/d_2$

$$\therefore \quad \frac{C_1}{C_2} = \frac{\varepsilon_o A/d_1}{\varepsilon_o A/d_2}$$

that is $C_1/C_2 = d_1/d_2$

$$\therefore \quad d_2 = d_1 \times C_1/C_2$$

$$= 1.5 \times 10^{-3} \times {}^{0.02}/0.0143 \qquad = \underline{\mathbf{2.1\ mm}}$$

(iv)

$$Q_1 = Q_2$$
$$= C_1 V_1$$
$$= 0.02 \times 100 \qquad = \underline{\mathbf{2\ \mu F}}$$

OR $Q_1 = Q_2$
$$= C_2 V_2$$
$$= 0.0143 \times 140 \qquad = \underline{\mathbf{2\ \mu F}}$$

Example 4.12

A parallel plate air dielectric capacitor consists of 21 plates, each of area 4 in². The plates are equally spaced 0.1 in between adjacent pairs. Calculate

    (a) the capacitance of this arrangement

    (b) the maximum d.c. voltage to which the capacitor may be charged if the electric strength of each dielectric must not exceed 15 kV/cm

    (c) the voltage across the capacitor if the dielectric flux density between each pair of plates is 5 μC/m²

Solution

(a) The capacitance of the arrangement is

$$C_T = \frac{\varepsilon_o \varepsilon_r (n - 1) A}{d}$$

where $\varepsilon_o = 8.85 \times 10^{-12}$

$$\varepsilon_r = 1 \text{ (air dielectric)}$$

$$n = 21$$

(contd)

(a)

$$(n-1) = (21-1) = 20$$

$$A = (2.54^2 \times 4) \times 10^{-4} \ m^2$$

$$d = (0.1 \times 2.54) \times 10^{-2} \ m$$

$$C_T = \frac{8.85 \times 10^{-12} \times 20 \times 2.54^2 \times 4 \times 10^{-4}}{0.1 \times 2.54 \times 10^{-2}}$$

$$= \frac{8.85 \times 20 \times 2.54^2 \times 4 \times 10^{-14} \times 10^{12} \ pF}{0.1 \times 2.54}$$

$$= \frac{8.85 \times 20 \times 2.54^2 \times 4 \times 10^{-2}}{0.254} = \underline{180 \ pF}$$

(b)

Electric strength of each dielectric is

$$E = V/d$$

∴ Maximum charging voltage is

$$V = E \times d$$

$$= 15 \ kV/cm \ \times \ (0.1 \times 2.54) \ cm$$

$$= 15 \times 0.254 \qquad = \underline{\textbf{3.81 kV}}$$

(c)

For $(n-1)$ dielectrics

$$C_T = 180 \ pF$$

For one dielectric

$$C = C_T/(n-1)$$

$$= 180/20 \qquad = \underline{9 \ pF}$$

Flux density, $D = \dfrac{\underline{coulombs \ (Q)}}{area \ (A)}$

$$Q = D \times A$$

$$= 5 \times 10^{-6} \times (2.54^2 \times 4) \times 10^{-4} \ m^2$$

$$= 5 \times 10^{-6} \times 25.8 \times 10^{-4} \qquad = \underline{0.0129 \ \mu C}$$

Since twenty dielectrics are connected in parallel, the voltage across the capacitor is

$$V = \frac{\underline{quantity \ of \ charge \ (Q)}}{capacitance \ with \ one \ dielectric \ (C)}$$

$$= \frac{0.0129 \times 10^{-6}}{9 \times 10^{-12}}$$

$$= \frac{12.9 \times 10^{-9}}{9 \times 10^{-12}}$$

$$= 1.4333 \times 10^3 \qquad = \underline{\textbf{1.4333 kV}}$$

Example 4.13

(a)   A 2 μF capacitor in parallel with a 3 μF capacitor is connected in series with a 5 μF capacitor. Calculate the effective capacitance of the circuit.

(b)   If 200 V d.c. is applied to the above circuit, determine the p.d across the capacitor and the charge stored by each capacitor.

(c)   If a counterpart capacitance of 6.2 μF is required, show how the capacitors could be connected to acquire this value.

Solution

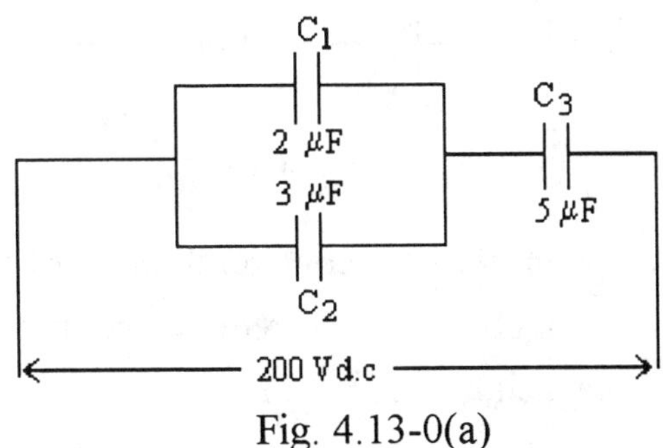

Fig. 4.13-0(a)

(a)   <u>Refer to Fig. 4.13-0(a)</u>

Equivalent capacitance of C_1 and C_2 in parallel is

$$C_{eq} = (3 + 2)\,\mu F = \underline{5\,\mu F}$$

The circuit is now reduced to that shown in Fig. 4.13-0(b).

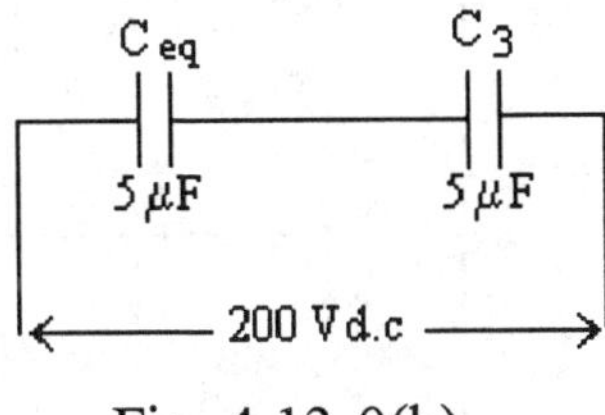

Fig. 4.13-0(b)

<u>Refer to Fig. 4.13-0(b)</u>

$$C_{eff} = \frac{C_{eq} \times C_3}{C_{eq} + C_3} = \frac{5 \times 5}{5 + 5} = \underline{\mathbf{2.5\,\mu F}}$$

(b)   Charge, $Q = C_{eff} \times V$

$$= 2.5 \times 10^{-6} \times 200 = \underline{500\,\mu F}$$

Since the parallel branch (2 μF and 3 μF) is in series with the 5 μF capacitor (C_3)

Charge Q on C_3 (5 μF) $= 500\,\mu C$

$\therefore$ P.d across C_3 (5 μF) $= Q/C = \dfrac{500 \times 10^{-6}}{5 \times 10^{-6}} = \underline{\mathbf{100\ V}}$

(contd)

P.d across parallel branch $= 200\ \text{Vd.c.} - 100\ \text{V}$ $\qquad$ = **100 V**

$\therefore$ Charge Q on C_1 (2 µF) = CV

$\qquad$ $= 2 \times 10^{-6} \times 100$ $\qquad$ = **200 µC**

Charge Q on C_2 (3 µF) $= 3 \times 10^{-6} \times 100$ $\qquad$ = **300 µC**

(c) The required connection is shown in Fig. 4.13-0(c).

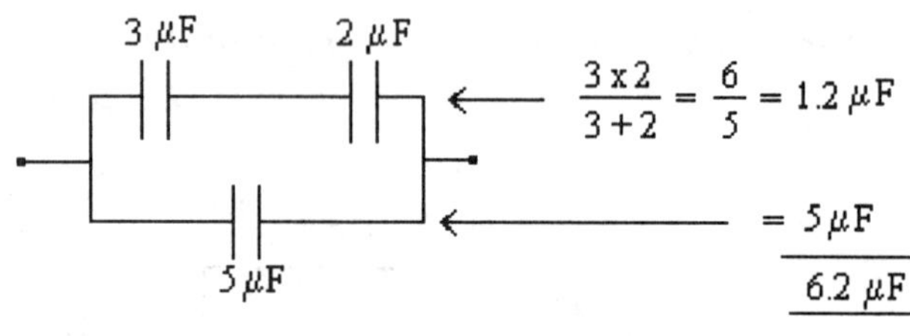

Fig. 4.13-0(c)

There are eight possible configurations, some of which may be eliminated by inspection. However, the required connection shown will give the effective capacitance of 6.2 µF.

Example 4.14

Determine the effective capacitance between points A and B shown in Fig. 4.14-0.

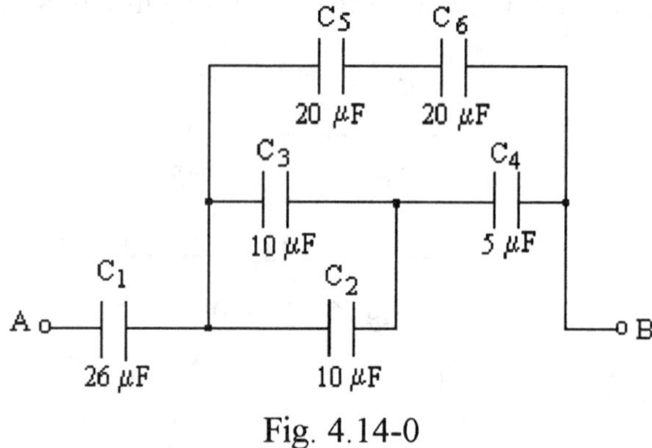

Fig. 4.14-0

Solution

The procedure to adopt in this type of problem is to apply successive simplification of the network as follows:

Reduce the series-connected C_5 and C_6 to an equivalent value. Thus

C_5 and C_6 in series $= 20 \times 20/(20 + 20)$

$= 400/40$ $\quad = $ 10 µF

Reduce the parallel combination C_2 and C_3 to an equivalent value. Thus

C_2 and C_3 in parallel $= 10 + 10$ $\quad = $ 20 µF

(contd)

The network of Fig. 4.14-0 is now reduced to that shown in Fig. 4.14-1.

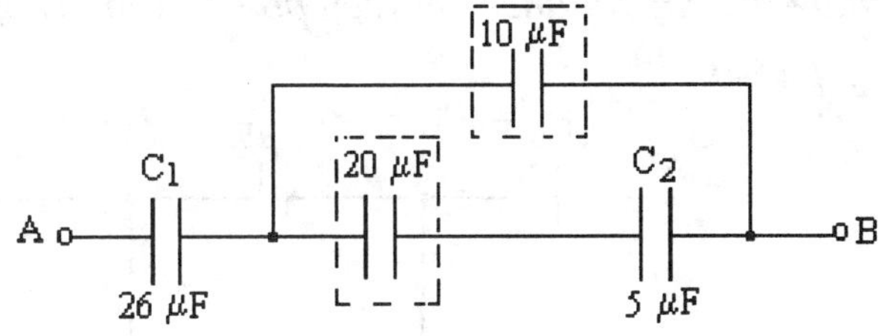

Fig. 4.14-1

<u>Refer to Fig. 4.14-1</u>

Combine the 20 µF and C_2 into a single unit value. Thus

20 µF and C_2 are in series $= \dfrac{20 \times 5}{20 + 5} = \underline{4\ \mu F}$

The network is now simplified and reduced to that shown in Fig. 4.14-2.

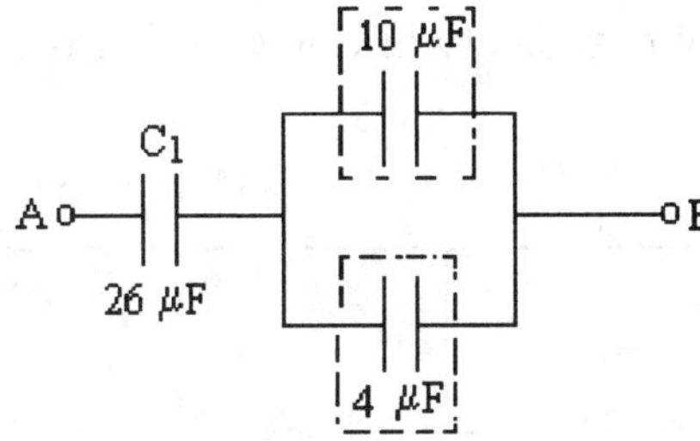

Fig. 4.14-2

At this stage the effective capacitance of the network may be calculated as follows:

$$C_{eff} = \dfrac{26 \times 14}{26 + 14} = \mathbf{\underline{9.1\ \mu F}}$$

However, if a further simplification is required, then combine the parallel branch of 10 µF and 4 µF; that is, counterpart value

$$= 10 + 4 = \underline{14\ \mu F}$$

The network can finally be reduced to that shown in Fig. 4.13-3.

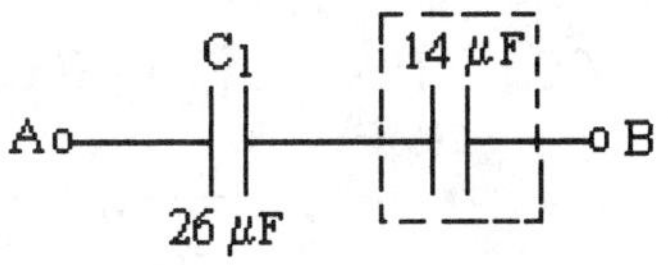

Fig. 4.14-3

<u>Refer to Fig. 4.14-3</u>

The two capacitors are in series. Therefore the effective capacitance of the network is

$$C_{eff} = \dfrac{26 \times 14}{26 + 14} = \mathbf{\underline{9.1\ \mu F}}$$

167

Example 4.15

Calculate the capacitance between points A and B of the network shown in Fig. 4.15-0

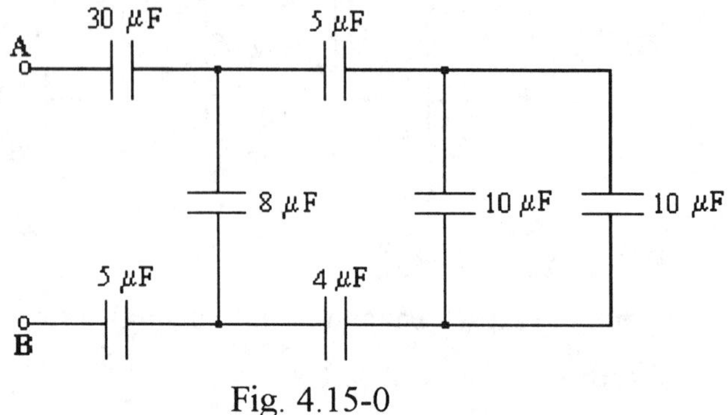

Fig. 4.15-0

Solution

<u>Refer to Fig. 4.15-1</u>

Start to simplify the network by working from **Right** to **Left.**

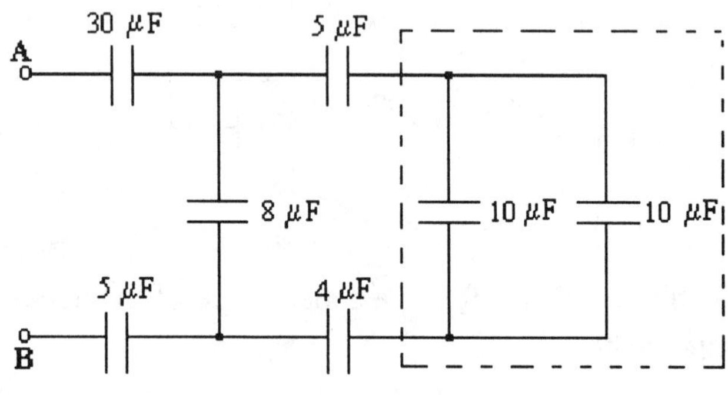

Fig. 4.15-1

(i)   Thus, the two 10 μF capacitors (enclosed in broken lines) are in parallel and their equivalent capacitance is

$$C_{P_1} \;\; = \; 10 + 10 \;\;\; = \; \underline{20 \; \mu F}$$

The circuit is now reduced to that of Fig. 4.15-2.

(ii)   Continue to simplify the network from **Right** to **Left.**

The 5 μF, 20 μF (C_{P_1}) and 4 μF capacitors are in series.

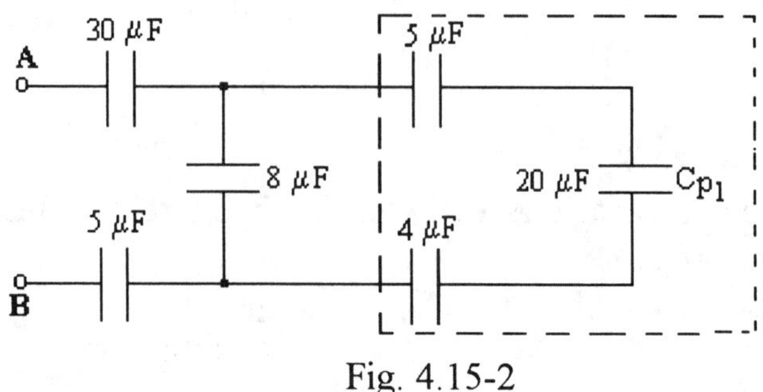

Fig. 4.15-2

168

(contd)

$$\therefore \quad 1/C_s = 1/5 + 1/4 + 1/20 = 10/20$$

$$\text{Inverting,} \quad C_s = 20/10 = \underline{2\,\mu F}$$

(iii)   The circuit is now reduced to that shown in Fig. 4.15-3.

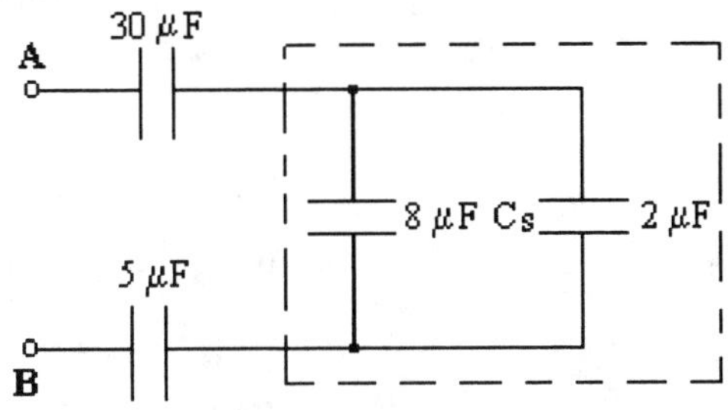

Fig. 4.15-3

The 8 μF capacitor and C_s (2 μF) are in parallel

$$\therefore \quad C_{P2} = 8 + 2 = \underline{10\mu F}$$

(iv)   The network is finally reduced to that shown in Fig. 4.15-4.

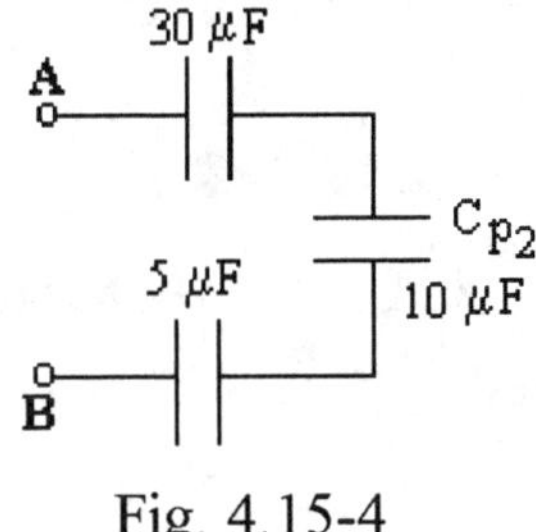

Fig. 4.15-4

The effective capacitance (C_{eff}) of the network between points A and B is

$$1/C_{eff} = 1/30 + 1/10 + 1/5 = 10/30$$

$$\therefore \quad C_{eff} = 30/10 = \mathbf{3\,\mu F}$$

Example 4.16

Two capacitors, 20 μF and 30 μF, are connected in series across a 1500 Vd.c. supply. Calculate (a) the p.d across each capacitor, (b) the charge on each capacitor, and (c) the capacitor with the greater stored energy.

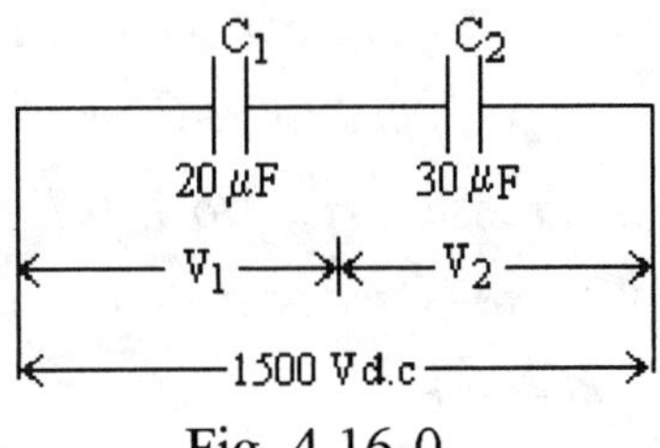

Fig. 4.16-0

Solution

(a) <u>Refer to Fig. 4.16-0</u>

Since the capacitors are connected in series, the same charging current flows into each capacitor. Therefore, the charge Q is identical for each.

$$\text{Hence,} \quad Q = C_1V_1 = C_2V_2$$

$$\therefore \quad C_1/C_2 = V_2/V_1$$

$$\text{that is,} \quad 20/30 = V_2/V_1$$

$$\therefore \quad 20V_1 = 30\,V_2$$

$$\text{Now} \quad V_2 = V_{\text{supply}} - V_1$$

$$\therefore \quad 20V_1 = 30\,[\,V_{\text{supply}} - V_1\,]$$

$$= 30\,[1500 - V_1]$$

$$50\,V_1 = 45000$$

$$V_1 = 45000/50 = \underline{\textbf{900 V}} \text{ (p.d across } C_1\text{)}$$

$$V_2 = 1500 - 900 = \underline{\textbf{600 V}} \text{ (p.d across } C_2\text{)}$$

(b) The charge on each capacitor is

$$Q = C_1V_1 = C_2V_2$$

$$= 20 \times 10^{-6} \times 900 \quad = \underline{\textbf{18000 }\mu\textbf{C}}$$

$$\underline{\textbf{or 0.018 C}}$$

$$\text{Charge in } C_1 = \text{Charge in } C_2 = \underline{\textbf{0.018 C}}$$

(c)
$$\text{Energy stored in } C_1 = \tfrac{1}{2}\,C_1V_1^2 \quad \text{joules}$$

$$= \tfrac{1}{2} \times 20 \times 10^{-6} \times 900^2$$

$$= 0.5 \times 20 \times 10^{-6} \times 81 \times 10^4$$

$$= 810 \times 10^{-2} \quad = \underline{\textbf{8.1 J}}$$

$$\text{Energy stored in } C_2 = \tfrac{1}{2} \times 30 \times 10^{-6} \times 600^2$$

$$= 0.5 \times 30 \times 10^{-6} \times 36 \times 10^4$$

$$= 540 \times 10^{-2} \quad = \underline{\textbf{5.4 J}}$$

It is evident from the above that the 20 μF capacitor has the greater energy stored.

Example 4.17

Two capacitors C_1 and C_2 are connected in series across a 240 Vd.c. supply. The p.d across C_1 is 130 V. When a 4 μF capacitor is connected across C_2, the p.d across C_1 rises to 140 V. Calculate (a) the capacitance of C_1 and C_2 and (b) the energy stored in each capacitor when the voltage across C_1 is 140 V.

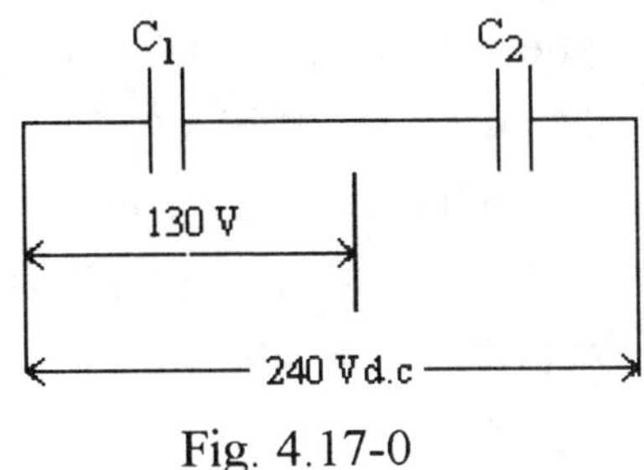

Fig. 4.17-0

Fig. 4.17-0 outlines the initial conditions.

<u>Refer to Fig. 4.17-0</u>

Since the capacitors are in series, the charge will be the same for either capacitor. That is

$$Q = C_1V_1 = C_2V_2$$
$$\therefore \quad V_1/V_2 = C_2/C_1$$

(a)   *Initial conditions*

$$V_2 = V_{supply} - V_1$$
$$= 240 - 130 = 110 \text{ V}$$
$$\therefore \quad {}^{130}/_{110} = C_2/C_1$$
$$130\,C_1 = 110\,C_2 \qquad \text{(Eq. 4.17-0)}$$

Final conditions

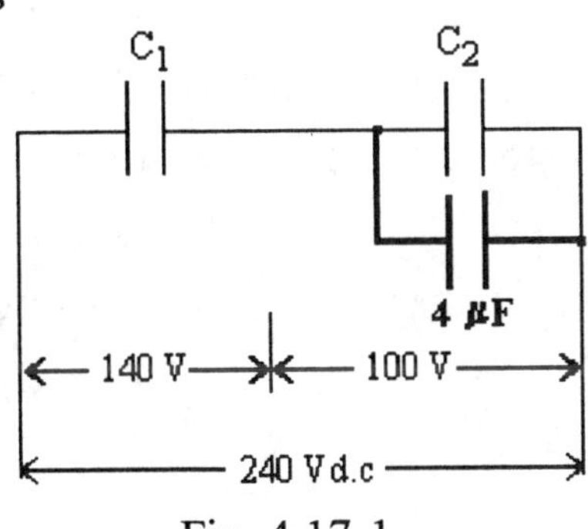

Fig. 4.17-1

<u>Refer to Fig. 4.17-1</u>

The 4 µF capacitor is connected across C_2. As a result V_1 across C_1 is increased to 140 V. Hence, V_2 across C_2 is 100 V.

$$\text{Now} \quad \frac{V_1}{V_2} = \frac{C_2 + 4}{C_1}$$
$$\frac{140}{100} = \frac{C_2 + 4}{C_1}$$
$$\therefore \quad 140\,C_1 = 100\,C_2 + 400 \qquad \text{(Eq. 4.17-1)}$$

$$\text{From Eq. 4.17-0} \quad C_2 = ({}^{130}/_{110})\,C_1$$

(contd)

Substituting this value of C_2 in Eq. 4.17-1, we get

$$140\,C_1 = 100 \times (^{130}/110)\,C_1 + 400$$

For simplicity, reduce by factor of 10. Hence

$$14\,C_1 = 10 \times (^{13}/11)\,C_1 + 40$$

$$14\,C_1 - (^{130}/11)\,C_1 = 40$$

$$154\,C_1 - 130\,C_1 = 440$$

$$24\,C_1 = 440$$

$$C_1 = {}^{440}/24 = \underline{\mathbf{18.33\ \mu F}}$$

Now substituting the value for C_1 in Eq.4.17-0, we get

$$C_2 = \frac{130 \times 18.33}{110} = \underline{\mathbf{21.66\ \mu F}}$$

(b)   Energy stored in each capacitor is as follows:

$$\text{Energy stored in } C_1 = \tfrac{1}{2}\,C_1 V_1^2 \quad \text{joules}$$

$$= \tfrac{1}{2} \times 18.33 \times 10^{-6} \times 140^2 \ \text{J}$$

$$= 0.5 \times 18.33 \times 10^{-6} \times 1.96 \times 10^4$$

$$= 17.96 \times 10^{-2} \qquad = \underline{\mathbf{0.1796\ J}}$$

$$\text{Energy stored in } C_2 = \tfrac{1}{2}\,C_2 V_2^2 \quad \text{joules}$$

$$= \tfrac{1}{2} \times 21.66 \times 10^{-6} \times 100^2$$

$$= 0.5 \times 21.66 \times 10^{-6} \times 1.00 \times 10^4$$

$$= 10.83 \times 10^{-2} \qquad = \underline{\mathbf{0.1083\ J}}$$

Energy stored in **4 μF** capacitor

$$= \tfrac{1}{2} \times 4 \times 10^{-6} \times 100^2$$

$$= 0.5 \times 4 \times 10^{-6} \times 1.00 \times 10^4$$

$$= 2.00 \times 10^{-2} \qquad = \underline{\mathbf{0.02\ J}}$$

Example 4.18

A capacitor of 30 μF is charged from a 500 Vd.c. supply and when fully charged is disconnected from the supply and immediately connected in parallel with another capacitor of 20 μF which is uncharged. Calculate

    (a)   *the energy stored in the 30 μF capacitor immediately before connection to the 20 μF capacitor*

    (b)   *the energy stored by the two capacitors immediately after being connected together*

    (c)   *the p.d across the combination*

Explain the possible source(s) of energy loss.

(a)   Energy stored by 30 µF capacitor when fully charged is

$$W = \tfrac{1}{2}CV^2 \text{ joules}$$
$$= \tfrac{1}{2} \times 30 \times 10^{-6} \times 500^2$$
$$= 15 \times 10^{-6} \times 25 \times 10^4 \qquad = \underline{\mathbf{3.75 \ J}}$$

(b)   Charge on 30 µF capacitor is

$$Q = CV$$
$$= 30 \times 10^{-6} \times 500 \qquad = \underline{0.015 \ C}$$

Since the capacitors are connected in parallel

$$C_{eff} = 30 + 20 \qquad = \underline{50 \ \mu F}$$

$$\text{Energy stored, } W = \tfrac{1}{2}CV^2$$
$$= \tfrac{1}{2}Q^2/C \qquad [V = Q/C]$$
$$= \tfrac{1}{2} \times (15 \times 10^{-3})^2 \times {}^1/C \ \ J$$
$$= \tfrac{1}{2} \times 225 \times 10^{-6} \times {}^1/(50 \times 10^{-6})$$
$$= \tfrac{1}{2} \times 225 \times 10^{-6} \times 10^6/50 \qquad = \underline{\mathbf{2.25 \ J}}$$

(c)   P.d across combination is

$$V = Q/C$$
$$= \frac{15 \times 10^{-3}}{50 \times 10^{-6}}$$
$$= 0.3 \times 10^3 \qquad = \underline{\mathbf{300 \ V}}$$

Possible sources of energy loss are

1.   radiation
2.   heating of the dielectrics
3.   copper loss in the connecting wires
4.   sparking which may occur when the connection is made

Example 4.19

Four perfect capacitors are connected to three terminals A, B and C as shown in Fig. 4.19-0.

     (a)   *Calculate the value of the capacitance that would be measured across terminals BC.*

     (b)   *Terminals B and C are now connected together. Determine the capacitance that would be measured across AB.*

     (c)   *Estimate the energy stored in the whole circuit in condition (b) above when 100 V is connected across AB.*

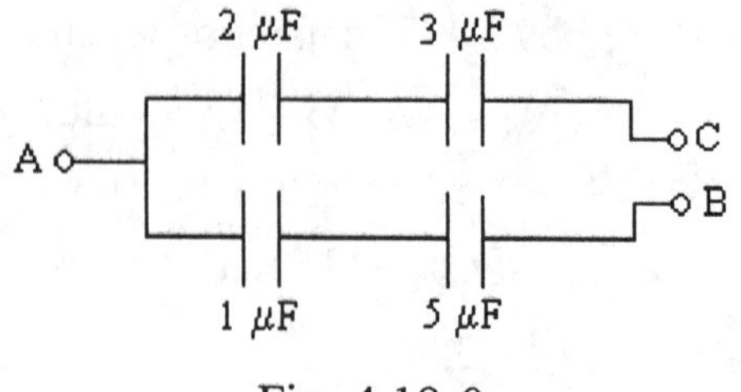

Fig. 4.19-0

Solution

(a)   <u>Refer to Fig. 4.19-0</u>

Across terminals BC, the four capacitors are in series.

$$\therefore \quad 1/C_{eff} = {}^1/2 + {}^1/3 + {}^1/1 + {}^1/5$$

$$= \frac{15 + 10 + 30 + 6}{30} = {}^{61}/30 \; \mu F$$

Inverting $C_{eff} = {}^{30}/61$ = **0.4918 μF**

(b)   When terminals B and C are connected together, the capacitors will be in *series-parallel.*

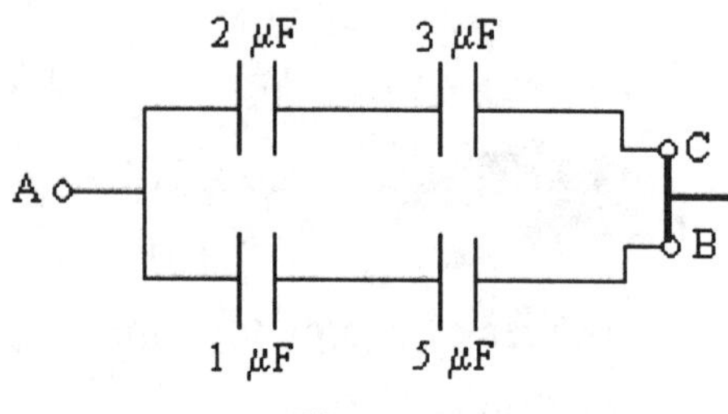

Fig. 4.19-1

<u>Refer to Fig. 4.19-1</u>

In series are   2 μF and 3 μF (along terminals AC)

and   1 μF and 5 μF (along terminals AB).

In parallel are terminals AC and AB [or branches AC and AB].

$$\therefore \quad C_{eq} = \text{series branch AC} + \text{series branch AB}$$

$$= \frac{2 \times 3}{2 + 3} + \frac{1 \times 5}{1 + 5}$$

$$= {}^6/5 + {}^5/6$$

$$= \frac{36 + 25}{30} = \frac{61}{30} = \mathbf{2.033 \; \mu F}$$

(c)   Energy stored, W $= \frac{1}{2} C_{eq} V^2$ joules

$$= \frac{1}{2} \times {}^{61}/30 \times 10^{-6} \times 100^2$$

$$= 1.0167 \times 10^{-2} \quad = \mathbf{0.0102 \; J}$$

Example 4.20

A 5 µF capacitor is connected to the terminals of an electrostatic voltmeter having negligible capacitance. A current-limiting resistance of 50 MΩ is arranged in series with the instrument and the combination is connected to a 500 V d.c. supply via a single-pole switch. Calculate the reading on the instrument 1 min after switching on to the supply.

Solution

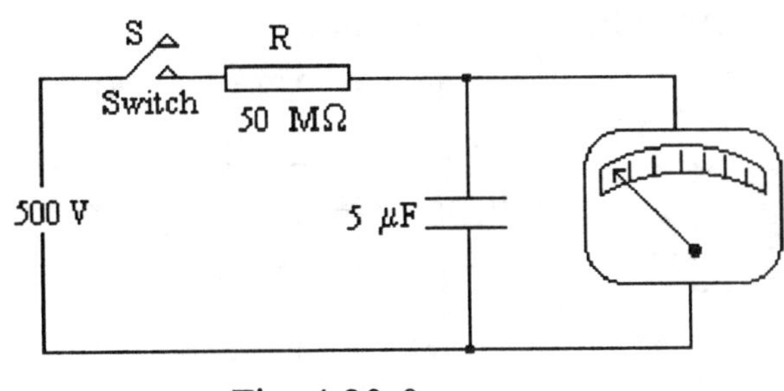

Fig. 4.20-0

<u>Refer to Fig. 4.20-0</u>

The charging current t sec after switching on is

$$i = \frac{V}{R} e^{-t/CR} \qquad [\text{Note: } t = 1 \text{ min} = 60 \text{ sec}]$$

$$= \frac{500 \; e^{-60/(5 \times 10^{-6} \times 50 \times 10^{6})}}{50 \times 10^{6}}$$

$$= 10 \times e^{-0.24} \times 10^{-6}$$

$$= 10 \times e^{-0.24} \; \mu A$$

$$= 10 \times \frac{1}{e^{\,0.24}}$$

$$= 10 \times \frac{1}{2.718^{\,0.24}}$$

$$= 10 \times 0.7866 \qquad\qquad = \underline{7.866 \; \mu A}$$

$$\therefore \quad \text{P.d across R} \;= i \times R$$

$$= 7.866 \times 10^{-6} \times 50 \times 10^{6} \qquad = \underline{393.3 \text{ V}}$$

$$\text{Reading on instrument} \;= 500 - 393.3 \qquad = \underline{\textbf{106.7 V}}$$

Example 4.21

A 20 µF capacitor is connected in series with a 50 kΩ resistor across a 250 Vd.c. supply. Calculate (a) the time taken for the voltage across the capacitor to rise to 150 V immediately after the supply is switched on and (b) the time taken for the voltage across the fully charged capacitor to fall to 50 V when it is discharged through a 1 MΩ resistor.

Solution

(a)   Using Eq. 4.0-13, we get

$$v = V\,(1 - e^{-t/CR})$$

$$150 = 250\,(1 - e^{-t/(20 \times 10^{-6} \times 50 \times 10^{3})})$$

$${}^{150}/_{250} = 1 - e^{-t/1}$$

$$0.6 = 1 - {}^{1}/_{e^{t}}$$

$$^{1}/_{e^{t}} = 1 - 0.6 \qquad = 0.4$$

$$e^{t} = {}^{1}/_{0.4} \qquad = 2.5$$

$$t\,\log_{e} e = \log_{e} 2.5$$

$$\therefore \qquad t = \underline{\textbf{0.916 sec}} = \underline{\textbf{time taken for }v\textbf{ to rise to 150 V}}$$

(b)

$$v = V e^{-t/CR}$$

$$50 = 250 e^{-t/(20 \times 10^{-6} \times 10^{6})}$$

$$^{50}/_{250} = e^{-t/20}$$

$$0.2 = {}^{1}/_{e^{0.05t}}$$

$$\therefore \qquad e^{0.05t} = {}^{1}/_{0.2} \ = 5$$

$$0.05t\,\log_{e} e = \log_{e} 5$$

$$\therefore \qquad t = {}^{1.61}/_{0.05} = \underline{\textbf{32.2 sec}}$$

$$= \textbf{time taken for }v\textbf{ to fall to 50 V}$$

Example 4.22

(a)   *Define the term "time constant" of a circuit that includes a resistor and a capacitor connected in series.*

(b)   *A capacitor of 100 µF is connected in series with a 15 kΩ resistor. Estimate the time constant of the circuit. If the combination is suddenly connected to a 250 Vd.c. supply, calculate*

       (i)   *the initial rate of rise of voltage across the capacitor*

       (ii)   *the initial charging current*

       (iii)   *the ultimate charge in the capacitor*

Solution

(a) The time constant T may be defined as the time required for the voltage across the capacitor to raise from 0 to 63.2% of its final value.

(b)
$$\text{Time constant, } T = CR \text{ sec}$$
$$= 100 \times 10^{-6} \times 15 \times 10^{3} \qquad = \underline{\textbf{1.5 sec}}$$

(i) The initial rate of rise of voltage across the capacitor is given by Eq. 4.0-13.

$$v = V(1 - e^{-t/CR})$$
$$= V - Ve^{-t/CR}$$
$$\therefore \quad \frac{dv}{dt} = \frac{Ve^{-t/CR}}{CR} \text{ volts/sec}$$

For the initial condition, $t = 0$.
$$\therefore \quad \frac{dv}{dt} = V/CR \text{ volts/sec}$$
$$= \frac{250}{100 \times 10^{-6} \times 15 \times 10^{3}} \text{ volts/sec}$$
$$= \frac{250 \times 10^{3}}{100 \times 15} \qquad = \underline{\textbf{166.67 V/sec}}$$

(ii) The initial charging current is given by Eq. 4.0-16.

$$i = Ie^{-t/CR}$$
$$= \frac{V e^{-t/CR}}{R}$$

For the initial condition, $t = 0$.
$$\therefore \quad i = V/R$$
$$= 250/(15 \times 10^{3}) \qquad = \underline{\textbf{16.67 mA}}$$

(iii) The ultimate charge is given by
$$Q = CV$$
$$= 100 \times 10^{-6} \times 250$$
$$= 25000 \times 10^{-6} \qquad = \underline{\textbf{0.025 C}}$$

Example 4.23

A 20 μF capacitor is charged to 100 V. If the capacitor is discharged through a wire of length 10 cm and diameter 0.5 mm weighing 0.15 g, estimate the temperature rise if all the stored energy of the capacitor is used in heating the wire. Take the specific heat of the wire to be 0.09 and 1 calorie = 4.18 J.

Solution

$$\text{Energy stored in capacitor} = \tfrac{1}{2} CV^2 \text{ joules}$$
$$= 0.5 \times 20 \times 10^{-6} \times 100^2 \qquad = \underline{0.1\ J}$$

This will be the amount of energy which will be converted to heat.

$$\therefore \quad \text{Heat produced} \quad = {}^{0.1}\!/4.18 \qquad\qquad = \underline{0.02392}\ \text{cal}$$

That is, mass x specific heat x temperature rise $t^{\circ}C$ $\qquad = 0.02392$ cal

$$\therefore \quad \text{Temperature rise, } t^{\circ}C = {}^{0.02392}\!/(0.15 \times 0.09) \qquad = \underline{\mathbf{1.772^{\circ}C}}$$

Example 4.24

A parallel plate capacitor is constructed from two circular metal discs separated by 3 layers of composite dielectric of equal thickness but varying relative permittivities of 3, 4 and 5, respectively. The diameter of each disc is 25.4 cm and the distance between them is 6 mm. Calculate the potential gradient and energy stored in each dielectric when 500 V is applied across the metal discs. Neglect fringing.

Solution

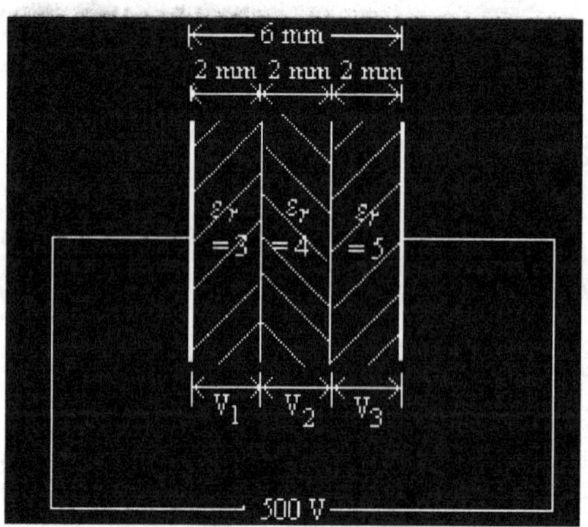

Fig. 4.24-0

Refer to Fig. 4.24-0, which gives a panoramic view of the problem.
Let E_1, E_2 and E_3 be the potential gradients in volts/meter of the dielectrics.

$$\text{Then} \quad (2E_1 + 2E_2 + 2E_3) \times 10^{-3} \qquad = 500\ V$$
$$\text{i.e.,} \quad 2(E_1 + E_2 + E_3) \qquad = 500 \times 10^3$$
$$\therefore \quad E_1 + E_2 + E_3 = \frac{500 \times 10^3}{2} \qquad = 250 \times 10^3\ V/m \qquad \text{(Eq. 4.24-0)}$$

(contd)

The flux density will be the same in each dielectric and is proportional to
to the potential gradient, that is

$$3\,kE_1 = 4\,kE_2 = 5\,kE_3$$

$$\therefore \quad 3\,E_1 = 4\,E_2 = 5\,E_3$$

$$\text{Hence} \quad E_1 = {}^5\!/_3\,E_3 \ \text{and} \ E_2 = {}^5\!/_4\,E_3$$

Substituting for E_1 and E_2 in Eq. 4.24-0, we get

$${}^5\!/_3\,E_3 + {}^5\!/_4\,E_3 + E_3 = 250 \times 10^3 \ \text{V/m}$$

$$20\,E_3 + 15\,E_3 + 12\,E_3 = 12\,(250 \times 10^3)$$

$$\therefore \quad E_3 = ({}^{12}\!/_{47}) \times 250 \times 10^3 = \underline{63.83 \ \text{kV/m}}$$

$$E_2 = {}^5\!/_4\,E_3 = ({}^5\!/_4) \times 63.83 = \underline{79.79 \ \text{kV/m}}$$

$$E_1 = {}^5\!/_3\,E_3 = ({}^5\!/_3) \times 63.83 = \underline{106.38 \ \text{kV/m}}$$

The voltages across the three dielectrics are now given by

$$V = d \times E$$

$$\therefore \quad V_1 = 2 \times 10^{-3} \times 106.38 \times 10^3 = \underline{\mathbf{212.76 \ V}}$$

$$V_2 = 2 \times 10^{-3} \times 79.79 \times 10^3 = \underline{\mathbf{159.58 \ V}}$$

$$V_3 = 2 \times 10^{-3} \times 63.83 \times 10^3 = \underline{\mathbf{127.66 \ V}}$$

<u>Refer to Eq. 4.0-3</u>

$$D\!/_E = Cd\!/_A$$

$$= \varepsilon_o \qquad (\text{where } Cd\!/_A = \varepsilon_o)$$

$$\therefore \text{ Flux density, } D = \varepsilon_o \varepsilon_r\,E$$

The flux density is the same for all three dielectrics. Therefore

$$\text{for } \varepsilon_r = 3, \quad D = \varepsilon_o \varepsilon_r\,E_1$$

$$= 8.85 \times 10^{-12} \times 3 \times 106.38 \times 10^3 \ \text{C/m}^2$$

$$= 8.85 \times 10^{-12} \times 3 \times 106.38 \times 10^9 \ \mu\text{C/m}^2$$

$$= 8.85 \times 3 \times 106.38 \times 10^{-3} = \underline{2.824 \ \mu\text{C/m}^2}$$

$$\text{Total flux in the dielectric} = Q\,A \qquad [A = \pi d^2\!/_4 \ \text{m}^2]$$

$$= 2.824\,[(\pi\!/_4) \times (25.4)^2 \times 10^{-4}]$$

$$= 2.824 \times (\pi\!/_4) \times 645.16 \times 10^{-4} = \underline{0.1431 \ \mu\text{C}}$$

$$\text{Energy stored , } W = \tfrac{1}{2}\,CV^2$$

$$= \tfrac{1}{2}\,Q/V \times V^2$$

$$= \tfrac{1}{2}\,Q\,V$$

Energy stored in the first dielectric is

$$= 0.5 \times 0.1431 \times 212.76 = \underline{\mathbf{15.22 \ \mu J}}$$

In the second dielectric $\quad = 0.5 \times 0.1431 \times 159.58 = \underline{\mathbf{11.42 \ \mu J}}$

In the third dielectric $\quad = 0.5 \times 0.1431 \times 127.66 = \underline{\mathbf{9.13 \ \mu J}}$

1.   A d.c. voltage of 150 V is applied to a capacitor whose capacitance is 120 μF. Calculate the quantity of electricity stored. [*Ans. 0.18 C*]

2.   An 80 μF capacitor holds 0.08 C of electricity. Determine the voltage across the capacitor.   [*Ans. 1000 V*]

3.   A capacitor is constituted of two flat metal plates, each of dimensions 30 cm x 20 cm, separated by Bakelite of uniform thickness 0.01 mm and relative permittivity 5.5. Calculate (a) the capacity of the capacitor and (b) the dielectric strength and flux density when the capacitor is charged to 500 V.   [*Ans. (a) 0.292 μF, (b) 50 kV/mm, 0.00243 C/m²*]

4.   Three capacitors having capacitances of 10 μF, 15 μF and 25 μF are connected in (a) series and (b) parallel. Calculate the total capacitance in each case.   [*Ans. (a) 4.84 μF, (b) 50 μF*]

5.   Three capacitors having capacitances of 20 μF, 30 μF and 60 μF are connected in series to an 800 Vd.c. supply. Calculate (a) the total capacitance, (b) the quantity of electricity stored in the combination and (c) the potential gradient across each capacitor.
        [*Ans. (a) 10 μF (b) 0.008 C (c) 400 V, 267 V, 133 V*]

6.   Show that when two capacitors C_1 and C_2 are connected in series the resultant capacitance is equal to the product of the capacitances divided by the sum of the capacitances.

7.   Two capacitors C_1 and C_2 are connected in series and their combined capacitance is 20 μF. If the C_1 is 25 μF, determine the value of C_2.
                                                        [*Ans. 100 μF*]

8.   Two capacitors of 60 μF and 40 μF are joined in parallel and this combination is connected in series with a 250 μF capacitor. A 700 Vd.c. supply is applied to this series-parallel combination. Determine the charge on the 60 μF capacitor.   [*Ans. 0.03 C*]

9.   Three capacitors, 2μF, 6μF and an unknown capacitor, are connected in series across a 240 Vd.c. supply. If the total capacitance of the three is 1 μF, calculate the value of the unknown capacitance and the voltage across each capacitor.   [*Ans. 3 μF, 120 V, 40 V, 80 V*]

10.   Calculate the capacitance between points A and B in the circuit shown in Fig. 4.P-10.   [*Ans. 8 μF*]

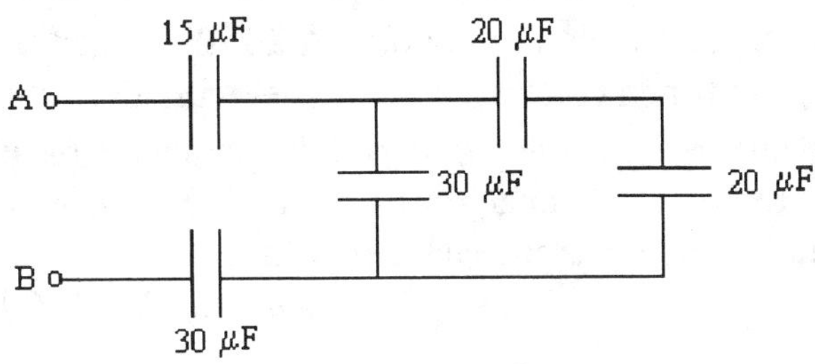

Fig. 4.P-10

11. A capacitor has two metal plates separated by the dielectric mica. The mica is 2 mm thick and has a relative permittivity of 6.5, and area of the plates is 400 cm². Calculate (a) the capacitance of the capacitor, (b) the electric stress in kV/m, (c) the flux density in μC/m² and (d) the charge in μC when 750 Vd.c. is applied across the plates.

[Ans. (a) 1152 pF, (b) 375 kV/m, (c) 21.6 μC/m², (d) 0.864 μC]

12. Two capacitors C_1 and C_2 are connected in series across 500 Vd.c. supply. C_1 has a capacitance of 200 pF and consists of two flat metal plates, each of area 100 cm², separated by paper dielectric 1 mm thick. Calculate (a) the relative permittivity of the paper dielectric, (b) the capacitance of C_2 if the electric strength in C_1 is 100 kV/m, (c) the total capacitance and (d) the electric flux density in C_1.

[Ans. (a) 2.26, (b) 50 pF, (c) 40 μF, (d) 2 μC/m²]

13. A capacitor consists of two flat metal plates separated by a dielectric of relative permittivity 2.5. The area of each plate is 200 cm². If the capacitance of the capacitor is 15 μF, determine the thickness of the dielectric.

[Ans. 2.21 x 10⁻⁵ mm]

14. Two parallel plate capacitors A and B, identical in construction except for the thickness of their dielectrics, are connected in series across a 220 Vd.c. supply. When the capacitors are fully charged, the p.d across A is 80 V. The capacitance of A is 0.01 μF, and the thickness of its dielectric is 2 mm. Calculate (a) the capacitance of B, (b) the total capacitance, (c) the thickness of the dielectric in B and (d) the charge in each capacitor.

[Ans. (a) 5710 pF, (b) 3640 pF, (c) 3.5 mm, (d) 0.8 μC]

15. Derive an expression for the effective capacitance when two capacitors are connected in series. A 3 μF capacitor is charged to a potential of 200 V and then connected in parallel with a 5 μF capacitor. Calculate the p.d across the combination. [Ans. 75 V]

16. Find an expression for the energy stored in a capacitor of capacitance C farads charged to a p.d of V volts. A 20 µF capacitor is charged to a p.d of 600 V and then connected in parallel with an uncharged capacitor of 28 µF. Calculate (a) the p.d across the parallel capacitors and (b) the energy stored in the capacitor before and after being connected in parallel. Account for the difference in energy loss.

[Ans. (a) 250 V (b) 3.6 J and 1.5 J]

17. Two capacitors of 2 µF and 3 µF are charged to 100 V and 200 V, respectively, and terminals of similar polarity are joined together. Calculate the energy stored before and after connecting in parallel. A 2 MΩ resistor is now connected across the two capacitors in parallel. Estimate the time taken for the capacitor voltage to fall to 50% of its initial value, using $v = Ve^{-t/CR}$

[Ans. Before 0.07 J, after 0.064 J; t = 6.93 sec]

18. The voltage applied to a 10 µF capacitor varies as follows: It rises uniformly for 0 to 500 V in 1.5 sec, remains constant at 500 V for 1 sec and then decreases uniformly to zero in 3.5 sec. Sketch a graph showing the variation of current during the 6 sec period, giving values for the charging and discharging current.   *[Ans. 3. 33 mA, 1.43 mA]*

19. The current flowing into a capacitor of 20 µF initially uncharged is maintained constant at 1 mA for 3 sec and then at 2 mA for 2 sec. For the following 3 sec it is discharged at 1.5 mA. Determine the p.d across the capacitor and the energy stored in it at the end of each of the three periods.

[Ans. 150 V, 350 V, 125 V, 0.225 J, 1.225 J, 0.156 J]

20. A capacitor of 100 µF is connected in series with an 8 kΩ resistor. Determine the time constant of the circuit. If 100 Vd.c. supply is suddenly connected to the combination, calculate (a) the initial rate of rise of voltage across the capacitor, (b) the initial charging current and (c) the ultimate charge in the capacitor.

[Ans. 0.8 sec; (a) 125 V/sec, (b) 12.5 mA, (c) 0.01 C]

21. A capacitor consists of two parallel metal plates, each of area 1600 cm², separated 6 mm apart by two composite dielectrics, mica and paper. The mica and paper dielectrics are 5 mm and 1 mm thick and have relative permittivities of 6 and 2 respectively. Calculate (a) the capacitance of the capacitor and (b) the potential gradient in each dielectric when a supply of 1.2 kV is applied across the plates.

[Ans. (a) 1062 pF (b) mica = 150 kV/m, paper = 450 kV/m]

22.   A capacitor consists of two parallel plates, each of area 3.77 m², separated by a 0.02 mm thick dielectric with a relative permittivity of 3. Determine (a) the charge on the capacitor when 100 Vd.c. is applied across the plates and (b) the voltage required to produce the same amount of charge on a second capacitor with plates of the same area but with a dielectric with twice the thickness and two-thirds the permittivity of the first capacitor. (c) Both capacitors are discharged and then connected in series. Calculate the charge taken from the 100 Vd.c. supply.

[Ans. (a) 500 μC, (b) 300 V, (c)125 μC]

23.   A parallel plate air-dielectric capacitor has a capacitance of 500 pF and a plate separation of 5 mm. Determine the area of each plate. If two composite dielectrics, each 2.5 mm thick, with relative permittivities 2 and 3 are inserted between the plates, calculate the value of the new capacitance.

[Ans. 0.283 m², 1200 pF]

24.   Two capacitors C_1 and C_2 are connected across a 100 Vd.c. supply. The voltages across the capacitors are 60 V and 40 V, respectively. A 2 μF capacitor is now connected in parallel with C_1 and the voltage across C_2 rises to 90 V. Calculate the capacitance of C_1 and C_2 in microfarads.

[Ans. C_1= 0.16 μF, C_2 = 0.24 μF]

CHAPTER 5

TRANSIENTS IN SIMPLE DIRECT CURRENT CIRCUITS

Introduction

When a circuit is switched from one state to another, there is a time duration before the circuit voltage or current settles to its steady state. This time duration is termed the *transient* response of the circuit. The output waveform so created may be in the shape of a ramp or a linearly rising form, a rectangular or square wave, or a spike or pulse as illustrated in Fig. 5.0-0.

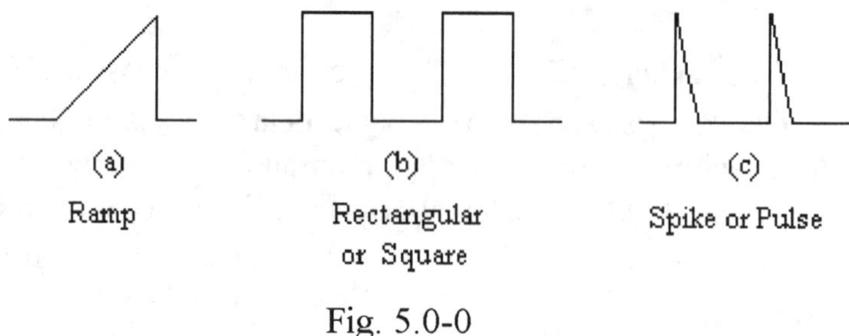

Fig. 5.0-0

Application of such waveforms has become increasingly prevalent in digital computers, radar, television and pulse-modulated forms of transmission systems.

R C Circuits

Charging

$$\text{Instantaneous current, } i = Ie^{-t/CR}$$

$$\text{Instantaneous p.d across C, } v_c = V(1 - e^{-t/CR})$$

Discharging

$$i = Ie^{-t/CR}$$

$$v_c = Ve^{-t/CR}$$

Note: CR is called the *time constant* of the circuit and is represented by the symbol τ. Hence, the instantaneous quantities may be written as follows:

$$i = Ie^{-t/\tau}$$

$$v = V(1 - e^{-t/\tau})$$

$\tau = CR$ sec   [Note: $\tau \equiv T = CR$ sec]

This is true for all equations involving the time constant CR.

R L Circuits

Charging

$$i = I(1 - e^{-Rt/L}) \text{ amperes}$$

Instantaneous current, $i = I(1 - e^{-Rt/L})$ amperes

Instantaneous p.d across R, $v = V(1 - e^{-Rt/L})$ volts

Discharging

$$i = Ie^{-Rt/L} \qquad \text{amperes}$$
$$v = Ve^{-Rt/L} \qquad \text{volts}$$

Note: $^L/R$ is called the *time constant* of the circuit, symbol τ.

Hence, the instantaneous values may be written as follows:

$$i = Ie^{-t/\tau}$$
$$v = V(1 - e^{-t/\tau})$$

$\tau = {}^L/R$ sec   [Note: $\tau \equiv T = CR$ sec)

This is true for all equations involving the time constant $^L/R$.

When $t = \tau$

$$i = Ie^{-t/\tau}$$
$$= Ie^{-1} \qquad = \underline{\textbf{0.368I amperes}}$$
$$v = V(1 - e^{-1})$$
$$= V(1 - 0.368) \qquad = \underline{\textbf{0.632V volts}}$$

Example 5.1

For the simple circuit shown in Fig. 5.1-0, sketch the graphs of the voltage across the resistor R, the voltage across the capacitor C and the charging current at any instant of time after switch S is closed. Hence derive the equations for

(i)   the charging current

(ii)   the voltage across capacitor C with switch S closed

(iii)   the voltage across C at any instant of time after S is opened

From the values obtained in (i), (ii) and (iii), explain the meaning of the term time constant denoted by τ.

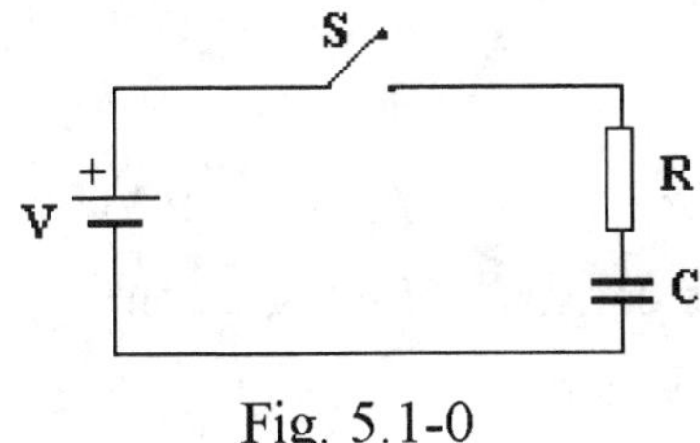

Fig. 5.1-0

185

With switch S closed

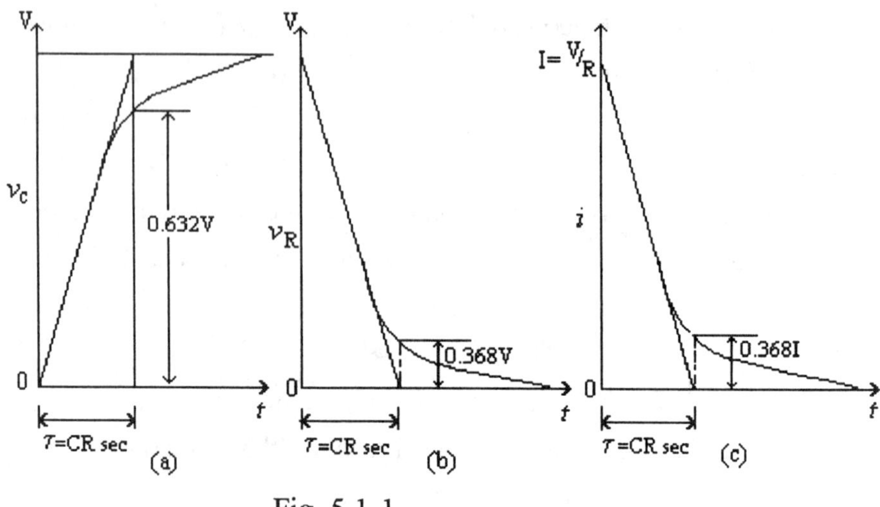

Fig. 5.1-1

<u>Refer to Fig. 5.1-1</u>

The graphs represent the following:

(a)   graph of voltage across C at any instant of time

(b)   graph of voltage across R at any instant of time

(c)   charging current in the circuit at any instant of time

(i)   The Charging Current

Total voltage V at any instant = voltage across R + voltage across C

That is,

$$V = iR + (1/C)\int i\,dt$$

$$dV/dt = R\,di/dt + i/C$$

Since the supply voltage V is constant

$$dV/dt = 0$$

$$\therefore \quad 0 = R\,di/dt + i/C$$

$$-R\,di/dt = i/C$$

$$di/dt = -i/CR$$

$$dt/di = -CR/i$$

$$dt = -CR\,di/i$$

$$\int dt = \int -CR\,di/i$$

$$\therefore \quad t = -CR\,\log_e i + C_1 \qquad\qquad \text{(Eq.5.1-0)}$$

(C_1 is called the constant of integration)

$$\text{When} \quad t = 0$$

$$i = V/R$$

(contd)

Substituting for t and i in Eq. 5.1-0, we get

$$0 = -CR \log_e (^V/R) + C_1$$
$$C_1 = CR \log_e (^V/R)$$

Substituting for C_1 in Eq. 5.1-0, we get

$$t = -CR \log_e i + CR \log_e (^V/R)$$
$$-t = CR \log_e i - CR \log_e (^V/R)$$
$$= CR (\log_e i - \log_e I)$$
$$\therefore \quad ^{-t}/CR = \log_e i - \log_e I$$
$$^{-t}/CR = \log_e {}^i/I$$
$$e^{-t/CR} = {}^i/I$$
$$\therefore \quad i = Ie^{-t/CR} \qquad \text{(Eq. 5.1-1)}$$

where R is in ohms, C is in farads and $t = \tau = CR$ is in seconds

(ii)   *Voltage across C with switch S closed*

$$\text{Voltage across} \quad R = V - v_c \quad (v_c = \text{voltage across C})$$
$$\therefore \quad i = \frac{V - v_c}{R}$$
$$\text{but} \quad i = Cdv/dt$$
$$\therefore \quad \frac{V - v_c}{R} = Cdv/dt$$
$$(^1/C)dt/dv = {}^R/(V - v_c)$$
$$(^1/CR)\, dt = {}^{dv}/(V - v_c)$$
$$\int(^1/CR)\, dt = \int^{dv}/(V - v_c)$$
$$(^1/CR)\, t = -\log_e (V - v_c) + C_1 \qquad \text{(Eq. 5.1-2)}$$

Again in this equation, C_1 is called the constant of integration.

$$\text{When} \quad t = 0, \quad v_c = 0$$

Substituting for t and v_c in Eq. 5.1-2, we get

$$0 = -\log_e V + C_1$$
$$C_1 = \log_e V$$

Substituting for C_1 in Eq. 5.1-2, we get

$$^t/CR = -\log_e (V - v_c) + \log_e V$$
$$-{}^t/CR = \log_e (V - v_c) - \log_e V$$
$$\therefore \quad Ve^{-t/CR} = V - v_c$$
$$v_c = V - Ve^{-t/CR}$$
$$\therefore \quad v_c = V(1 - e^{-t/CR}) \qquad \text{(Eq. 5.1-3)}$$

(contd)

<u>Refer to Eq. 5.1-3</u>

> V is the voltage across the capacitor C when $t = 0$
>
> R is the resistance in ohms
>
> C is the capacitance in farads

(iii)   *Voltage across C at any instant of time after switch S is opened*

$$\text{When} \quad t \;=\; 0$$

$$v_c \;=\; V_s \quad \text{(the supply voltage)}$$

$$\text{But} \quad V \;=\; 0 \quad \text{(the final voltage)}$$

Substituting for t, v_c and V in Eq. 5.1-2, we get

$$t/CR \;=\; -\log_e (V - v_c) + C_1 \qquad \text{(Refer to Eq. 5.1-2)}$$

$$\therefore \quad 0 \;=\; -\log_e (0 - V_s) + C_1$$

$$-C_1 \;=\; -\log_e (-V_s)$$

$$C_1 \;=\; \log_e (-V_s)$$

$$\therefore \quad t/CR \;=\; -\log_e (V - v_c) + \log_e (-V_s)$$

$$-t/CR \;=\; \log_e (V - v_c) - \log_e (-V_s)$$

$$\;=\; \log_e (-v_c) - \log_e (-V_s) \quad (V = 0)$$

$$\;=\; \log_e (^{-v}c/-V_s)$$

$$\;=\; \log_e (^{v}c/V_s)$$

$$\therefore \quad \mathbf{v_c \;=\; V_s e^{-t/CR}} \qquad \text{(Eq. 5.1-4)}$$

Time Constant

Equation 5.1-1 gives the current

$$i \;=\; Ie^{-t/CR}$$

$$\text{and} \quad \tau \;=\; CR \text{ sec}$$

Substituting for CR in Eq. 5.1-1, we get

$$i \;=\; Ie^{-t/\tau}$$

$$\;=\; Ie^{-1} \qquad (t = \tau)$$

$$\;=\; \mathbf{0.368I}$$

The time constant, denoted by τ, is the time taken for the current to fall to 0.368 (36.8%) of its original value. Refer to Fig. 5.1-1(c).

<u>Refer to Eq. 5.1-3</u>

$$v_c \;=\; V (1 - e^{-t/CR})$$

$$\;=\; V (1 - e^{-CR/CR}) \quad (t = CR)$$

$$\;=\; V (1 - 0.368)$$

$$\;=\; \mathbf{0.632\ V}$$

(contd)

<u>Reference to Eq. 5.1-3</u>

$$v_c = 0.632 \text{ V}$$

That is, the time constant τ is the time taken for the voltage to rise to
0.632 or 63.2% of its final value. Refer to Fig. 5.1-1(a).

<u>Refer to Eq. 5.1-4</u>

$$v_c = V_s e^{-t/CR}$$
$$= V_s e^{-1} \qquad (t = CR \text{ sec})$$
$$= \mathbf{0.368V_s}$$

This is the case of the decay of voltage across the capacitor C and
hence defines the time constant τ as the time taken for the voltage
to fall to 0.368 or 36.8% of its original value.

Example 5.2

(a) *The circuit of Fig. 5.2-0(a) is a simple L R circuit connected to a*
d.c. supply through a single-pole switch S. Explain with the aid of
sketches the circuit behavior when switch S is closed.

(b) *Derive expressions for*
   (i) *the rise of current in the circuit*
   (ii) *the decay of current in the circuit*

(c) *The components of an L R circuit are shown in Fig. 5.2-0(b).*
Calculate (i) *the value of current after 0.5 sec*
   (ii) *the time constant of the circuit*
   (iii) *the time taken for the current to reach 1.0 A*

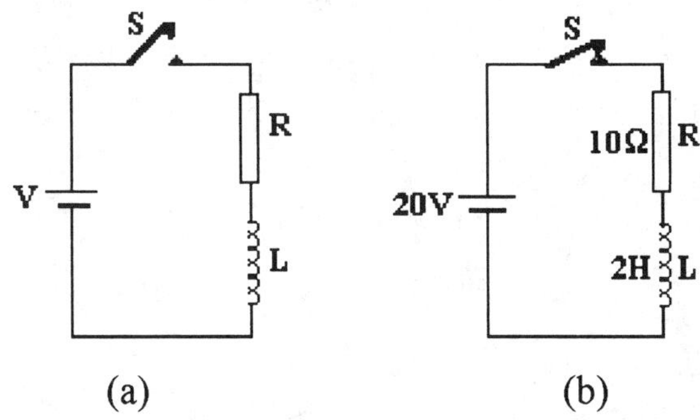

Fig. 5.2-0

Solution

(a)   <u>Consider Fig.5.2-0(a)</u>

At the instant of closing switch S, the current does not immediately rise to its final value, $I = V/R$.

$$\text{Let} \quad t \quad = \quad \text{time after closing switch S}$$
$$\text{and} \quad i \quad = \quad \text{instantaneous current}$$
$$\text{Now} \quad V \quad = \quad L\,di/dt + Ri$$
$$\therefore \quad V - L\,di/dt \quad = \quad Ri$$

On closing switch S, $i \quad = \quad 0$

$$\therefore \quad Ri \quad = \quad 0$$
$$\text{and} \quad V \quad = \quad L\,di/dt \quad \text{or} \quad di/dt = V/L$$

This is the initial rate of change of current.

When the current I has reached its final steady value

$$di/dt \quad = \quad 0 \qquad \text{and} \qquad V = IR$$
$$\text{Final value of} \quad I \quad = \quad V/R$$

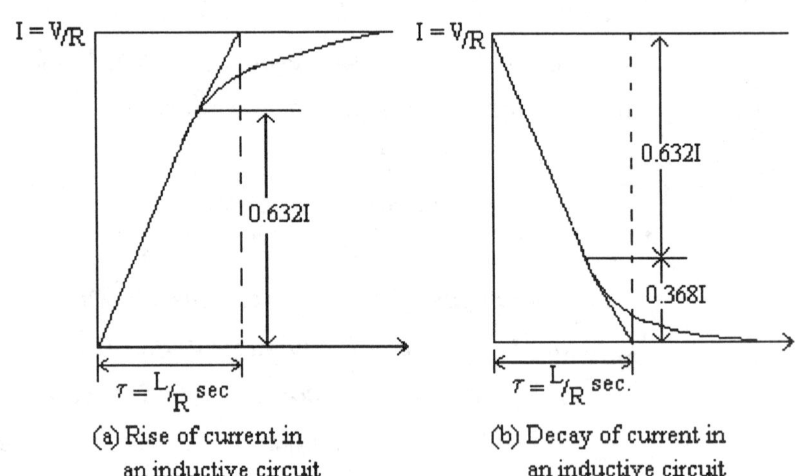

Fig. 5.2-1

(b) (i) Rise of current in an inductive circuit. Refer to Fig. 5.2-1(a)

The equation for the current is

$$L\,di/dt + Ri \quad = V$$
$$L\,di/dt \quad = V - Ri$$
$$di/dt \quad = (V - Ri)/L$$

Inverting,
$$dt/di \quad = L/(V - Ri)$$
$$\int dt \quad = L\int \frac{1}{(V - Ri)}\,di$$
$$t \quad = -(L/R)\log_e (V - Ri) + C$$

(contd)

$$\text{When } t = 0, \quad i = 0$$

$$\therefore \quad C = (L/R) \log_e V$$

$$\text{Substituting for C}, \quad t = -(L/R) \log_e (V - Ri) + (L/R) \log_e V$$

$$\text{i.e.,} \quad t = (L/R) \log_e [V/(V - Ri)]$$

$$Rt/L = \log_e [V/(V - Ri)]$$

$$\therefore \quad V/(V - Ri) = e^{Rt/L}$$

$$\text{Inverting,} \quad (V - Ri)/V = e^{-Rt/L}$$

$$V - Ri = Ve^{-Rt/L}$$

$$Ri = V - Ve^{-Rt/L}$$

$$i = (V - Ve^{-Rt/L})/R$$

$$= (V/R)(1 - e^{-Rt/L})$$

$$\therefore \quad i = I(1 - e^{-Rt/L}) \quad (V/R = I) \quad (\text{Eq. 5.2-1})$$

Refer to Eq. 5.2-1

The time constant τ for an inductive circuit is L/R sec. This is the time taken for the current to rise to 0.632 or 63.2% of its final value.

Refer to Fig. 5.2-1(a)

$$i = I(1 - e^{-R/L \times L/R}) \quad (t = \tau = L/R)$$

$$= I(1 - e^{-1})$$

$$= \mathbf{0.632I}$$

(ii) *Decay of current in an inductive circuit*

Refer to Fig. 5.2-1(b), the decay curve.

From Eq. 5.2-0, determine the value of C.

$$\text{That is,} \quad Rt/L = -\log_e (I - i) + C$$

$$\text{When } t = 0, \quad i = V/R$$

$$\text{Final value of} \quad I = 0$$

Substituting these values in Eq. 5.2-0, we get

$$0 = -\log_e (0 - V/R) + C$$

$$\text{from which} \quad C = \log_e (-V/R)$$

$$\therefore \quad Rt/L = -\log_e (I - i) + \log_e (^-V/R)$$

But final value of $I = 0$

$$\therefore \quad Rt/L = -\log_e (-i) + \log_e (^-V/R)$$

$$= \log_e [i/(V/R)]$$

$$\therefore \quad e^{-Rt/L} = i/(V/R)$$

$$\therefore \quad i = \mathbf{Ie^{-Rt/L}} \quad (\text{Eq. 5.2-2})$$

(contd)

In the case of the decay of current, the time constant τ of an inductive circuit is $^L/R$ sec and is the time taken for the current to fall to 0.368 or 36.8% of its initial value.

$$i = Ie^{-Rt/L}$$
$$= Ie^{-1} \qquad (t = \tau = {}^L/R)$$
$$\mathbf{i = 0.368I}$$

(c) *(i)*

$$i = I(1 - e^{-Rt/L})$$
$$i = {}^V/R = {}^{20}/10 = \underline{2\,A}$$
$$\therefore \quad i = 2[1 - e^{-(10 \times 0.5)/2}]$$
$$= 2[1 - e^{-2.5}]$$
$$= 2[1 - 0.0821] \qquad = \underline{\mathbf{1.836\,A}}$$

(ii)

Time constant, $\tau \quad = {}^L/R$ sec
$$= {}^2/10 \qquad\qquad = \underline{\mathbf{0.2\,sec}}$$

(iii)

$$i = I(1 - e^{-Rt/L})$$
$$= 2(1 - e^{-5t})$$

Given that $i = 1$ A it follows that

$$1 = 2 - 2e^{-5t}$$
$$2e^{-5t} = 2 - 1$$
$$e^{-5t} = {}^1/2$$
$${}^1/e^{5t} = {}^1/2$$
$$e^{5t} = 2$$
$$5t = \log_e 2$$
$$= 0.6931$$
$$\therefore \quad t = {}^{0.6931}/5 \qquad = \underline{\mathbf{0.1386\,sec}}$$

Example 5.3

(a) *A symmetrical square wave of frequency 1.0 kHz and amplitude +10V is applied to the input terminals of the circuits shown in Fig. 5.3-0(a) and Fig. 5.3-0(b). For each condition, sketch on the same time scale two cycles of the input and output waveforms. State which circuit may be used as:*

    (i)   *an integrator*

    (ii)   *a differentiator*

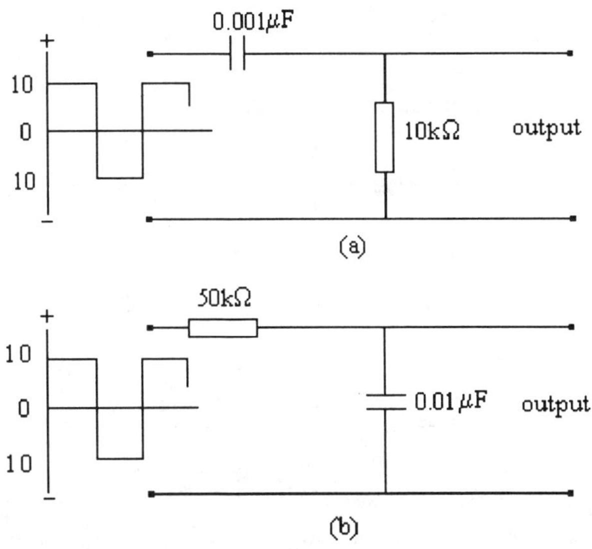

Fig. 5.3-0

(b)   Determine the ratio of the capacitive reactance(X_c) to the resistance(R) in each circuit assuming a sinusoidal waveform of 1.0 kHz. Comment on your results.

Solution

(a)   <u>Refer to Fig. 5.3-0(a)</u>

Time constant, τ = CR sec

$\qquad$ = $0.001 \times 10^{-6} \times 10 \times 10^{3}$ = 0.01 msec

<u>Refer to Fig. 5.3-0(b)</u>

Time constant, τ = CR sec

$\qquad$ = $0.01 \times 10^{-6} \times 50 \times 10^{3}$ = 0.5 msec

Periodic time of input signal = $^{1}/f$ = $^{1}/1000$ = <u>1 msec</u>

(a)   *(i)*   The circuit of Fig. 5.3-0(a) produces an output voltage proportional to the slope of the input signal voltage. Therefore, the circuit is called a **differentiator.**

The time constant of the circuit is very small (0.01 ms or 10 µs); compare with the periodic time of the input signal (1.0 ms). The output waveform is almost ideal if it was not for the width of the spikes. Ideally, the spikes would have zero width.

(contd)

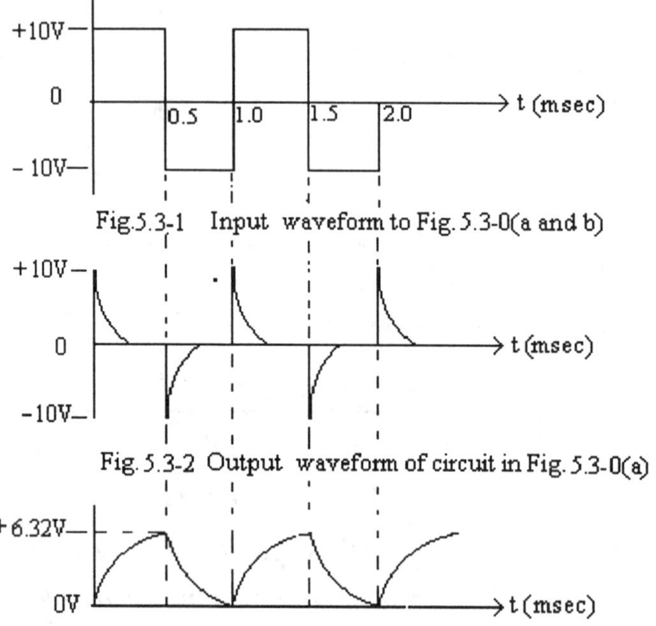

Fig.5.3-1   Input waveform to Fig. 5.3-0(a and b)

Fig. 5.3-2 Output waveform of circuit in Fig. 5.3-0(a)

Fig. 5.3-3   Output waveform of circuit in Fig. 5.3-0(b)

(a)   *(ii)*   The circuit of Fig. 5.3-0(b) is an **integrator** and it should produce an output proportional to the area under the input signal waveform, that is, a sawtooth waveform from the square wave input. Unfortunately, this is not the ideal output waveform due to the fact that the time constant of the circuit (0.5 ms) corresponds to the time of the input signal.

For a better sawtooth waveform from the square wave input, the time constant of the **integrator** circuit should be approximately 5 times this value.

(contd)

(b)
<u>Refer to Fig. 5.3-0(a)</u>. *The differentiator circuit*

The reactance, X_c $= {}^1/(2\pi f\, C)$

$= 10^6/(2 \times 3.14 \times 1000 \times 0.001)$

$\approx 159\ k\Omega$

$\therefore$ Ratio of $X_c : R$ $= 159 : 10$ $\qquad \approx \mathbf{\underline{16:1}}$

<u>Refer to Fig. 5.3-0(b)</u>. *The integrator circuit*

The reactance, X_c $= {}^1/(2\pi f\, C)$

$= 10^6/(2 \times 3.14 \times 1000 \times 0.01)$

$\approx 15.9\ k\Omega$

$\therefore$ Ratio of $X_c : R$ $= 15.9 : 50$ $\qquad \approx \underline{1:3}$

The reactance of the capacitor when a sinusoidal signal is applied to the *differentiator circuit* [Fig. 5.3-0(a)] is approximately 16 times the value of R.

That is

$$X_c \gg R$$

Ratio $\qquad$ 16 : 1

Similarly, the reactance of the capacitor in the *integrator circuit* [Fig. 5.3-0(b)] is about $^1/3$ the value of R.

That is

$$X_c \ll R$$

Ratio $\qquad$ 1 : 3

Example 5.4

A circuit is connected as shown in Fig. 5.4-0. The switch S was at position A for a long time. It is then switched to position B for 750 μsec and then back to position A. Calculate the output voltage v_o for the time duration between which the switch was toggled. Sketch a shape of the waveform.

Solution

<u>Refer to Fig. 5.4-0</u>

Since the switch S was closed in position **A** for a long time, it is assumed that the capacitor is fully charged. Therefore

$$v_o \quad = V_o \ = 250\ V$$

(contd)

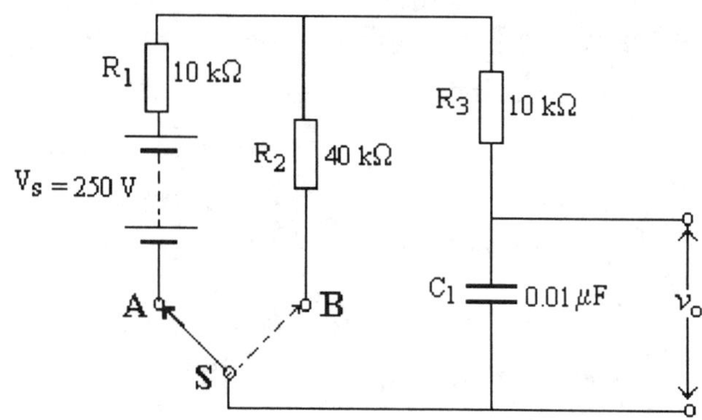

Fig. 5.4-0

At $t = 0$, switch S is toggled to position **B** and the capacitor immediately starts to discharge for a time period of 750 μsec.

$$\therefore \quad i = Ie^{-t/CR}$$

$$I = {}^{V}/(R_2 + R_3)$$

$$= {}^{250}/[(40 + 10) \times 10^3] \quad = \underline{5 \text{ mA}}$$

$$CR = 0.01 \times 10^{-6} \times 50 \times 10^3 \quad = \underline{500 \text{ μS}}$$

$$v_o = -iR_3 + v_c$$

$$= -(5 \times 10^{-3} \times 10 \times 10^3) + 250$$

$$= -50 + 250 = \underline{200 \text{ V}}$$

Therefore the voltage v_o at the instant of switching is

$$v_o = 200e^{-750 \times 10^{-6}/500 \times 10^{-6}}$$

$$= 200e^{-1.5}$$

$$= 200 \times {}^{1}/e^{1.5}$$

$$= 200 \times {}^{1}/4.482$$

$$= 200 \times 0.2231 \quad = \underline{\textbf{44.62 V}}$$

The shape of the output waveform is shown in Fig. 5.4-1

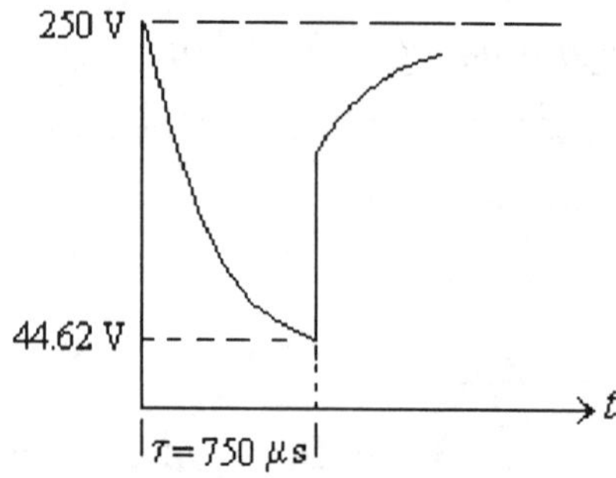

Fig. 5.4-1. Output waveform

Example 5.5

Determine the output voltage v_o shown in the circuit of Fig. 5.5-0 under the following conditions:

    (a)   the instant switch S is closed

    (b)   at a time 5 sec later

    (c)   after a long period

Determine the initial rate of change of the voltage across the 2 MΩ resistor and the final charge on capacitor C in microcoulombs. The capacitor is initially uncharged.

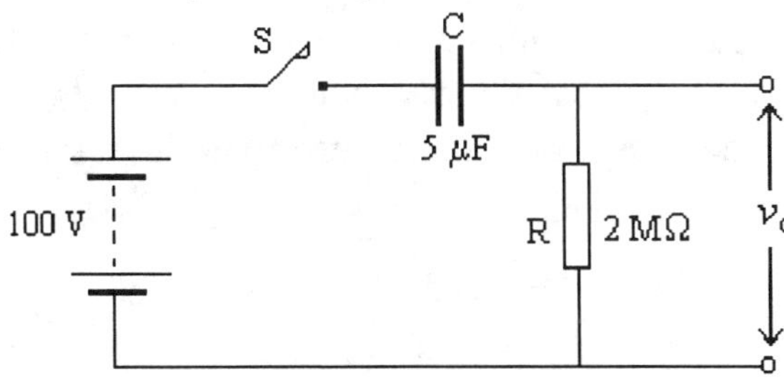

Fig. 5.5-0

Solution

(a)   At the instant switch S is closed

$$v_o = \mathbf{100\ V}$$

(b)   5 sec later, $v_o = Ve^{-t/CR}$

$$= 100\ e^{-5/(5 \times 10^{-6} \times 2 \times 10^{6})}$$
$$= 100e^{-\frac{1}{2}}$$
$$= 100 \times \frac{1}{e^{\frac{1}{2}}}$$
$$= 100 \times 0.6065 \qquad = \mathbf{60.65\ V}$$

(c)   After a long period, $v_o = \mathbf{0\ V}$

Initial rate of change of voltage across R is

$$v = Ve^{-t/CR}$$
$$\frac{dv}{dt} = -\frac{Ve^{-t/CR}}{CR}$$

At $t = 0$, $\dfrac{dv}{dt} = -\dfrac{V}{CR}$

$$= -\frac{100}{(5 \times 2)} \qquad = -\mathbf{10\ V/sec}$$

Final charge on C is

$$Q = CV$$
$$= 5 \times 10^{-6} \times 100\ \text{Coulombs}$$
$$= 5 \times 10^{-6} \times 100 \times 10^{6}\ \mu C$$
$$= 5 \times 100 \qquad = \mathbf{500\ \mu C}$$

1. An 8 μF capacitor is connected in series with a 0.5 MΩ resistor to a 300 V d.c. supply. Construct a curve showing the growth of voltage across the capacitor and determine the time taken for the p.d across the capacitor to rise to 240 V.   [*Ans. 6.44 sec*]

2. A capacitor of 8 μF and a resistor of 500 kΩ are connected in series across a 300 Vd.c. supply. The voltage across the capacitor reaches 240 V in 6.44 sec. Calculate the current which will then be flowing in the resistor.
                    [*Ans. 120 μA*]

3. A 16 μF capacitor is connected in series with a 100 kΩ resistor. The combination is connected via a switch to a 200 Vd.c. supply. Calculate the p.d across the capacitor and the current in the resistor 2.4 sec after switching on.   [*Ans.155.4 V, 0.446 mA*]

4. A field winding of a d.c. machine has an inductance of 60 H and a resistance of 30 Ω. A discharge resistor of 50 Ω is permanently connected in parallel with the winding which is excited from a 200 V supply. Construct a curve showing the decay of current in the field winding with a time base from its steady value after the supply was switched off, and from the curve derive the value of this current 0.6 sec after the supply has been switched off.   [*Ans. 3 A (by calculation)*]

5. A coil of 2000 turns has a resistance of 50 Ω and produces a magnetic flux of 5 mWb when a steady current of 4 A is passed through it. The supply to the coil is 200 Vd.c. Calculate (a) the time constant of the coil, (b) the induced e.m.f in the coil at the instant the current has risen to 3 A and (c) the rate of growth of current at this same instant.
                    [*Ans. (a) 0.05 sec, (b) 50 V, (c) 20 A/sec*]

6. A constant voltage V is maintained across an inductance of L henrys in series with a resistance R Ω. Write an expression for the current that flows in the circuit t seconds after switching on. A relay coil of resistance 200 Ω and inductance 8 H is connected in series with a 100 Ω resistor and a 60 V battery. The relay operates when the current in its coil is 31.6 mA. Calculate the time taken for the relay to operate.   [*Ans. 4.68 msec*]

7. A coil has a self-inductance of 0.75 H and a resistance of 25 Ω. It is connected to a d.c. supply of 50 V. If the supply is suddenly removed and replaced by a short circuit, calculate the time taken for the current to decay from 1.5 A to 0.5 A.   [*Ans. 33 msec*]

8.   A 4 μF capacitor is connected in series with a 1 MΩ resistor across a
250 Vd.c. supply. Use an approximate method to construct the current-
time curve for the charging period. Hence, determine the instantaneous
current after a time equal to the time constant of the circuit. Compare this value
with that obtained by calculation.   [*Ans. 0.092 mA (by calculation)*]

CHAPTER 6

DIRECT CURRENT GENERATORS

Introduction

The e.m.f E generated by a d.c. generator is given by the expression

$$E = \frac{2\,\Phi\,Z\,N\,p}{60\,c} \qquad \text{(Eq. 6.0-0)}$$

where Φ = flux per pole (Wb)

       Z = number of conductors in the armature

       N = speed in revolutions/minute (r.p.m)

       p = number of pairs of poles

       c = number of parallel paths through windings between positive and negative brushes

         = 2 for wave winding

         = 2p for lap winding (twice the number of pairs of poles)

$\therefore$ Z/c gives the number of conductors in series between the positive and negative brushes.

In Eq. 6.0-0 only the flux Φ and the speed N are variables for a given machine.

$$\therefore \quad E \propto \Phi\, N$$

$$\text{or} \quad E = k\,\Phi\, N$$

where $k = \dfrac{2\,Z\,p}{60\,c}$

The armature winding has a resistance R_a ohms.

When the armature is delivering current I_a, the terminal voltage V_T is less than the generated e.m.f E because of the voltage drop in the armature. That is

    Terminal voltage = generated e.m.f – voltage drop

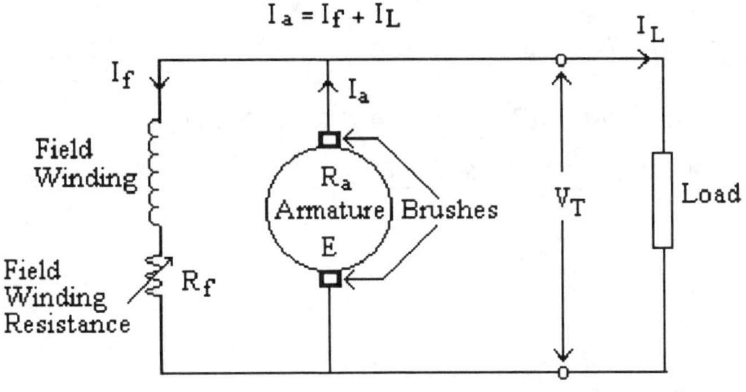

Fig. 6.0-0

<u>Refer to Fig. 6.0-0</u>

In general, if

$$E = \text{e.m.f generated in armature}$$
$$I_a = \text{armature current}$$
$$R_a = \text{armature resistance}$$
$$V_T = \text{terminal voltage}$$

then $\quad V_T = E - I_a R_a \qquad\qquad$ (Eq. 6.0-1)

or $\quad E = V_T + I_a R_a$

Methods of Excitation

This section describes the methods used to connect the field and armature windings.

Technically, there are two methods of excitation:

1. Separately excited and
2. Self-excited

Self-excited is subdivided into three groups:

(i) Shunt excited

(ii) Series excited

(iii) Compound excited

A brief explanation of the above now follows.

Separately excited generator:

<u>Refer to Fig. 6.0-1</u>

The field winding is connected to a separate source of external supply

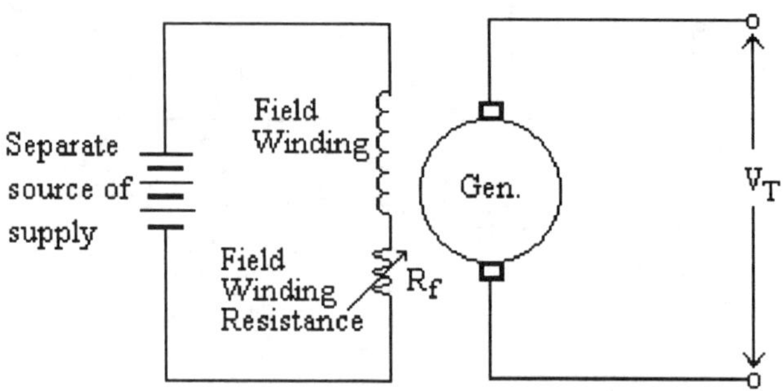

Fig. 6.0-1. Separately excited generator

Shunt excited generator:

<u>Refer to Fig. 6.0-2</u>

The field winding is connected in shunt or parallel with the armature winding.

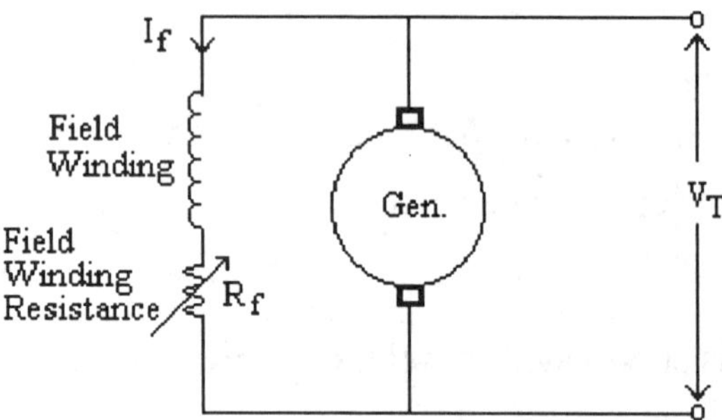

Fig. 6.0-2. Shunt-excited generator

Series-excited generator:

<u>Refer to Fig. 6.0-3</u>

The field winding is connected in series with the armature winding. It is of interest to note that the series-excited generator will not excite on NO-LOAD, the reason being that current will only flow in the series winding when a load is connected across the terminals, V_T, enabling effective excitation . Also, excitation may not be effective if the load resistance is very high at starting.

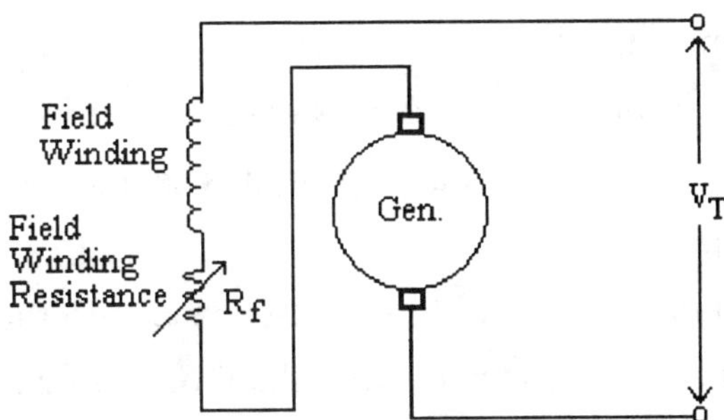

Fig. 6.0-3. Series-excited generator

Compound excited generator:

<u>Refer to Fig. 6.0-4</u>

This is a combination of the shunt and series field windings.

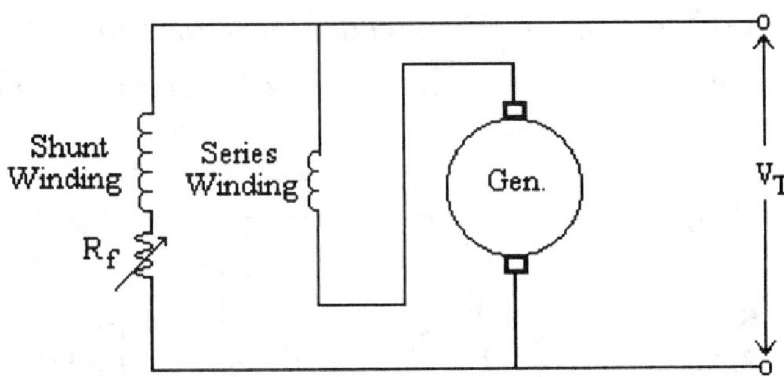

Fig. 6.0-4. Compound excited generator

The compound excited generator may operate as

– under compound
– level compound
– over compound

Each of the above describes the number of turns required in the series field winding to compensate for the fall of terminal voltage with increase of load.

Notes:

> When the shunt winding is connected in series with the series field winding as in Fig. 6.0-4, the generator is said to be connected "long shunt".
>
> When the shunt winding is connected directly across the armature of the compound generator, it is said to be connected "short shunt".
>
> In practice, it is of little consequence whether the shunt field winding is connected long shunt or short shunt, the reason being that the shunt current is negligibly small in comparison with the full-load current, and the number of turns in the series field winding is very small compared with the number of turns in the shunt field winding.

Example 6.1

A 24 kW, 240 Vd.c. shunt generator shown in Fig. 6.1-0 has a field winding resistance of 60 Ω and an armature resistance of 0.1 Ω. Calculate the value of the e.m.f generated when the machine is on full load.

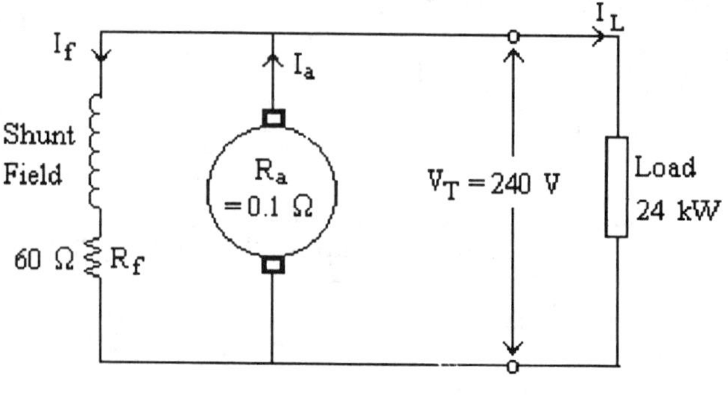

Fig. 6.1-0

Solution

<u>Refer to Fig. 6.1-0</u>

Let I_f = field current

R_f = field resistance

R_a = armature resistance

I_a = armature current

Full-load current, $I_L = \dfrac{kW}{V_T}$ $= \dfrac{24 \times 10^3}{240}$ = <u>100 A</u>

Field current, $I_f = \dfrac{V_T}{R_f}$ $= \dfrac{240}{60}$ = <u>4 A</u>

Armature current, $I_a = I_L + I_f$ $= 100 + 4$ = <u>104 A</u>

E.m.f generated, E = terminal voltage + voltage drop in armature

$= V_T + I_a R_a$

$= 240 + (104 \times 0.1)$

$= 240 + 10.4$ = <u>**250.4 V**</u>

Example 6.2

A shunt-connected d.c. generator has an armature resistance of 0.1 Ω and a field winding resistance of 100 Ω. If the shunt field current is 2 A, find the value of the armature e.m.f when the load resistance is 5 ohms.

Solution

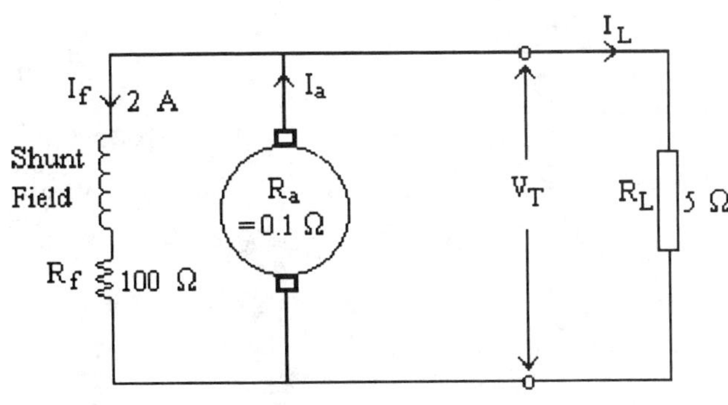

Fig. 6.2-0

(contd)

<u>Refer to Fig. 6.2-0</u>

The voltage across the shunt winding, V_{shunt}, is the same as the terminal voltage V_T. That is

$$
\begin{aligned}
V_T \quad &= V_{shunt} \\
&= I_f R_f \\
&= 2 \times 100 \qquad = \underline{200\ V}
\end{aligned}
$$

$$
\begin{aligned}
\text{Load current,} \quad I_L \quad &= V_T/R_L \\
&= 200/5 \qquad = \underline{40\ A}
\end{aligned}
$$

$$
\begin{aligned}
\text{Armature current,} \quad I_a \quad &= I_L + I_f \\
&= 40 + 2 \qquad = \underline{42\ A}
\end{aligned}
$$

$$
\begin{aligned}
\therefore \quad \text{Armature e.m.f,} \quad E \quad &= V_T + I_a R_a \\
&= 200 + (42 \times 0.1) \\
&= 200 + 4.2 \qquad = \underline{\mathbf{204.2\ V}}
\end{aligned}
$$

Example 6.3

An eight-pole wave-wound armature of a d.c. generator having 49 slots with 10 conductors per slot is driven at 1000 r.p.m. Calculate the e.m.f generated in the armature if the useful flux per pole is 15 mWb.

Solution

<u>Using Eq. 6.0-0</u>

$$
E = \frac{2\,\Phi\,Z\,N\,p}{60\,c}
$$

where

$$
\begin{aligned}
\Phi &= 15 \times 10^{-3}\ Wb = 0.015\ Wb \\
Z &= 49 \times 10 \quad = 490\ \text{conductors} \\
N &= 1000\ \text{r.p.m} = 1000/60\ \text{rev/sec} \\
c &= 2\ \text{for wave-wound armature} \\
p &= 4\ \text{pairs}
\end{aligned}
$$

$$
\begin{aligned}
\therefore \quad E &= \frac{2 \times 0.015 \times 490 \times 1000 \times 4}{60 \times 2} \\
&= \frac{58800}{60 \times 2} \qquad = \underline{\mathbf{490\ V}}
\end{aligned}
$$

Example 6.4

A four-pole lap-wound armature of a d.c. generator is driven at 500 r.p.m and it is required to generate 250 V with a useful flux of 0.05 Wb per pole. Assuming an armature with 120 slots, determine the number of conductors per slot.

Solution

Using Eq. 6.0-0

$$E = \frac{2\,\Phi\,Z\,N\,p}{60\,c}$$

where   flux, Φ = 0.05 Wb

Z = unknown quantity

N = 500 r.p.m

p = 2 pairs

c = 4 (number of poles for lap wound)

E = 250 V

Number of conductors, Z $= \dfrac{E \times 60\,c}{2\,\Phi\,N\,p}$

$$= \frac{250 \times 60 \times 4}{2 \times 0.05 \times 500 \times 2}$$

$$= \frac{60 \times 10^3}{100} \qquad = \underline{600 \text{ conductors}}$$

Since there are 120 slots in the armature

$\therefore$ Conductors per slot $= {}^{Z}\!/_{120} = {}^{600}\!/_{120} = \underline{5}$

Example 6.5

(a)   *The e.m.f E generated in an armature conductor is given by the expression*

$$E = \frac{2\,\Phi\,Z\,N\,p}{60\,c}$$

where Φ *= useful flux per pole in webers entering or leaving the armature*

Z *= total number of conductors in armature*

N *= speed in revolutions per minute (r.p.m)*

p *= number of pairs of poles*

c *= number of parallel paths*

Show the derivation of this expression.

(contd)

(b)   *A four pole d.c. shunt generator has a lap-wound armature with 1200 conductors. If the flux per pole is 0.02 Wb, determine the speed at which the armature must be driven to generate 460 V.*

Solution
(a)

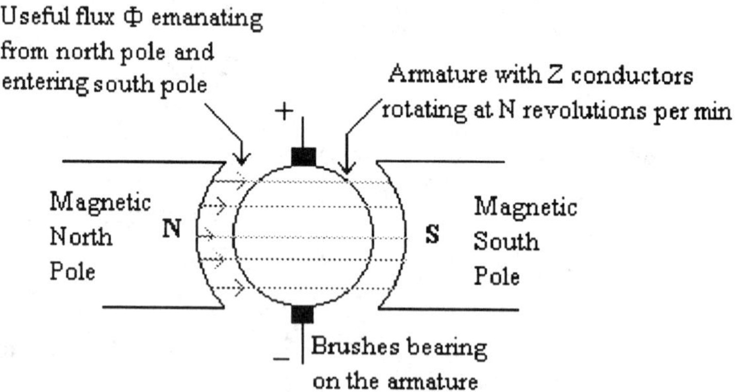

Fig. 6.5-0. A two-pole d.c. generator

Consider the two-pole d.c. generator shown in Fig. 6.5-0.

Let Φ = useful flux per pole entering and leaving the armature

        p = number of pairs of poles

        N = speed in revolutions per minute

Time taken to complete one revolution $= \dfrac{60}{N}$ sec

Time taken to rotate 1 pole pitch $= \dfrac{60/N}{2p}$ sec

$$= \dfrac{60}{2Np} \text{ sec}$$

Average rate at which flux is cut by the rotating armature while moving 1 pole pitch; that is, flux cut per second

$$= \dfrac{\Phi}{60/2Np} \text{ Wb/sec}$$

$$= \dfrac{2\Phi Np}{60} \text{ Wb/sec}$$

$\therefore$ Average e.m.f generated in each conductor is

    Average e.m.f $= \dfrac{2\Phi Np}{60}$ V

(contd)

Let   Z   = total number of conductors in the armature

       c   = the number of parallel paths through windings between positive and negative brushes

The armature may be *wave wound* or *lap wound*.

For *wave-wound armature*,   c   = 2

For *lap-wound armature*,   c   = number of poles in the generator

∴ Conductors in series per path   = Z/c

Hence, total e.m.f between brushes = average e.m.f in each conductor x number of conductors in series per path

$$= \frac{2\,\Phi N p}{60} \times Z/c$$

That is, $E = \dfrac{2\,\Phi Z N p}{60\,c}$ V

(b)   The e.m.f generated, $E = \dfrac{2\,\Phi Z N p}{60\,c}$

∴ $N = \dfrac{E \times 60\,c}{2\,\Phi Z p}$

where   E   = 460 V

       c   = 4   (lap-wound armature with 4 poles)

       Φ   = 0.02 Wb

       p   = 2 pairs

∴ $N = \dfrac{460 \times 60 \times 4}{2 \times 0.02 \times 1200 \times 2}$

$$= \frac{110400}{96} \qquad = \mathbf{1150\ r.p.m}$$

Example 6.6

A d.c. shunt generator has a generated e.m.f of 480 V on no-load. The armature has a resistance of 0.1 Ω and a field resistance of 200 Ω. The voltage drop across each brush is 1.0 V. Calculate what percentage increase in speed will be required to maintain a terminal p.d of 480 V when delivering a load current of 150 A.

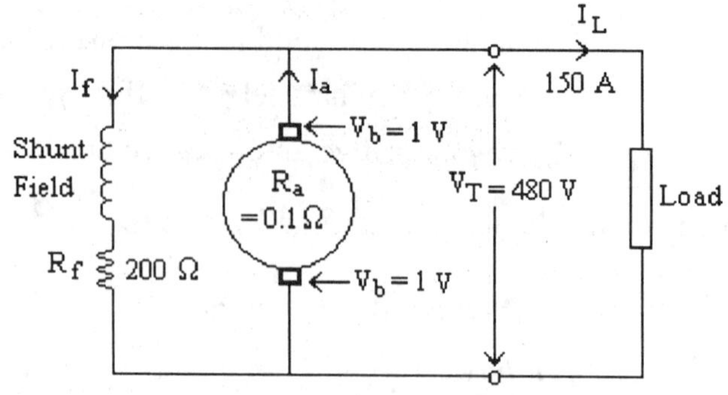

Fig. 6.6-0

<u>Refer to Fig. 6.6-0</u>

$$\text{Armature current, } I_a = I_L + I_f$$
$$= 150 + (^{480}/200)$$
$$= 150 + 2.4 \qquad\qquad = \underline{152.4 \text{ A}}$$

$$\text{Generated e.m.f, } E = V_T + (I_a R_a) + V_{\text{brushes}}$$
$$= 480 + (152.4 \times 0.1) + 2$$
$$= 480 + 15.24 + 2 \qquad = \underline{497.24 \text{ V}}$$

$$\text{The no-load voltage} = 480 + (2V_b) \qquad = \underline{482 \text{ V}}$$

Now $E \propto \Phi N$ and for a given value of Φ, $E \propto N$.

∴   Percentage increase in speed is

$$= \frac{497.24 - 482}{482} \times 100\% \qquad = \underline{\mathbf{3.16\%}}$$

Example 6.7

A d.c. shunt-excited generator driven at 600 r.p.m has the following magnetization characteristics:

Field current (I_f)	1	2	3	4	5	6	7	8	A
Generated e.m.f (E)	23	45	67	85	100	112	121	126	V

Calculate the load current when

    (a)   the terminal p.d is 120 V, the field resistance is 15 Ω and the speed is 600 r.p.m

    (b)   the terminal p.d is 144 V, the field resistance is 18 Ω and the speed is 700 r.p.m

The armature resistance is 0.02 Ω and the effects of armature reaction may be neglected.

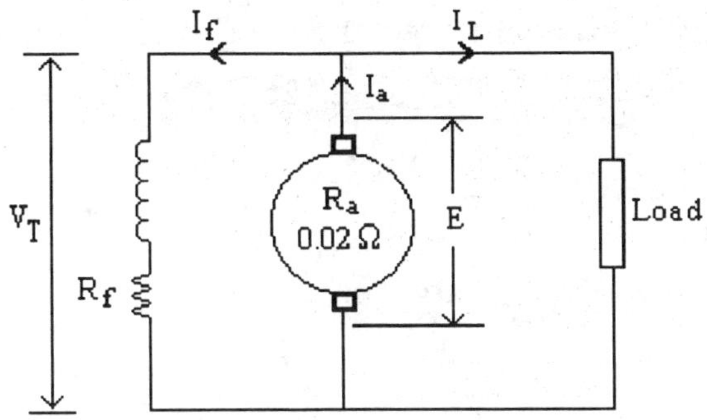

Fig. 6.7-0

<u>Refer to Fig. 6.7-0</u>

$$E \quad = \text{generated e.m.f}$$
$$R_a \quad = \text{armature resistance}$$
$$I_L \quad = \text{load current}$$
$$I_f \quad = \text{shunt field current}$$
$$R_f \quad = \text{shunt field resistance}$$
$$I_a \quad = I_L + I_f$$
$$V_T \quad = \text{terminal p.d} = E - I_a R_a$$

(a) Field current, $\quad I_f \quad = V_T/R_f$

$$= {}^{120}/_{15} \qquad = \underline{8 \text{ A}}$$

From the magnetization characteristics

$$I_f \quad = 8 \text{ A}, E = 126 \text{ V}$$

Now $\quad V_T \quad = E - I_a R_a$

$\therefore \quad I_a \quad = (E - V_T)/R_a$

where $\quad E \quad = 126 \text{ V}$
$$V_T \quad = 120 \text{ V}$$
$$R_a \quad = 0.02 \ \Omega$$

$\therefore \quad I_a \quad = (126 - 120)/_{0.02} \quad = \underline{300 \text{ A}}$

Load current, $\quad I_L \quad = I_a - I_f$

$$= 300 - 8 \qquad = \underline{\textbf{292 A}}$$

(contd)

(b) From *(a)* field current, $I_f = V_T/R_f = {}^{144}/18 = \underline{8\ A}$

At 600 r.p.m this field current gives a generated e.m.f of

$$E = 126\ V$$

Now $\quad E \propto N$

$\therefore \quad E = 126 \times {}^{700}/600 = \underline{147\ V}$

The voltage drop in the armature resistance is

$$I_a R_a = 147 - 144 = \underline{3\ V}$$

$\therefore \quad I_a = {}^{3}/R_a = {}^{3}/0.02 = \underline{150\ A}$

Hence, load current, $\quad I_L = I_a - I_f$

$$= 150 - 8 = \underline{\mathbf{142\ A}}$$

Example 6.8

A shunt generator is to be converted to a level compound generator by the addition of a series field winding. From a test conducted on the generator, the results are as follows:

No-load

shunt field current $\quad = 6.5\ A$

terminal voltage $\quad = 480\ V$

Full-load

shunt field current $\quad = 8.25\ A$

terminal voltage $\quad = 480\ V$

load current $\quad = 200\ A$

The shunt winding has 1500 turns per pole. Determine the number of series turns required per pole.

Solution

Ampere-turns per pole required on no-load is

$$= I_f N$$

$$= 6.5 \times 1500 = \underline{9750\ AT}$$

Ampere-turns per pole required on full-load is

$$= I_f N$$

$$= 8.25 \times 1500 = \underline{12375\ AT}$$

(contd)

Ampere-turns per pole to be provided by series winding is

$$= 12375 - 9750 \qquad = \underline{2625\ AT}$$

Armature current at full-load $= 200 + 8.25 \qquad = \underline{208.25\ A}$

Number of series turns per pole (at full-load current) is

$$= 2625/208.25 \qquad = \underline{12.61\ turns}$$
$$= \mathbf{\underline{13\ turns}}$$

(Poles cannot be wound with a fraction of turns)

Example 6.9

It is desired to level compound a shunt generator so that the terminal p.d at full load of 200 A is equal to that at no-load. When connected as a shunt generator, the field current had to be increased from 4.1 A to 5.8 A in order to maintain the terminal p.d constant between no-load and full-load. If the generator is to be connected long shunt and the turns per pole on the shunt field is 1200, calculate the series turns per pole required (neglect voltage drop in the series field winding).

After its conversion, the above generator had a terminal voltage of 240 V at full-load current and the resistances of the armature, shunt and series fields were 0.12 Ω, 80 Ω and 0.08 Ω, respectively. Calculate the value of the e.m.f being generated under these conditions.

Solution

Change in field current to maintain the terminal voltage constant is

$$= 5.8 - 4.1 \qquad = \underline{1.7\ A}$$

$\therefore$ Change in exciting ampere-turns per pole is

$$= 1.7 \times 1200 \qquad = \underline{2040\ AT}$$

Armature current with full-load is

$$= 200 + 5.8 \qquad = \underline{205.8\ A}$$

$\therefore$ Required number of series turns per pole is

$$= 2040/205.8 \qquad = \underline{9.9\ turns}$$
$$= \mathbf{\underline{10\ turns}}$$

Generated e.m.f, E $= V_T + I_a\,(R_a + R_{se})$

$$= 240 + 205.8\,(0.12 + 0.08)$$
$$= 240 + 24.696 + 16.464$$
$$= \mathbf{\underline{281.16\ V}}$$

Example 6.10

A shunt generator which has the following details: $R_a = 0.15\ \Omega$, $R_f = 150\ \Omega$, *generated e.m.f = 300 V. The generator supplies a load of 200 A. Calculate the terminal voltage of the generator. Hence show that the field current may be neglected.*

A second generator with the same armature resistance and shunt field resistance but with a generated e.m.f of 291 V is now connected in parallel with the first generator. Determine the current supplied by each generator and the new terminal voltage. State any assumptions made.

Solution

$$\text{Field current,}\quad I_f = V_T/R_f$$
$$= V_T/150$$

$$\text{Terminal voltage,}\ V_T = E - I_a R_a$$
$$= E - (I_L + I_f)\, R_a$$
$$= 300 - [200 + (V_T/150)]\, 0.15 \qquad \text{(Eq. 6.10-0)}$$
$$= 300 - (30 + 0.001\, V_T)$$

<u>Refer to Eq. 6.10-0</u>

$$V_T \quad < 300$$
$$\therefore \qquad V_T/150 \quad < 2A\ < I_L/100$$
$$\text{That is,} \qquad I_L \quad \gg V_T/150$$

Therefore, the field current may be neglected.

$$\text{Hence} \qquad V_T \ = 300 - 30 \qquad\qquad = \underline{\mathbf{270\ V}}$$

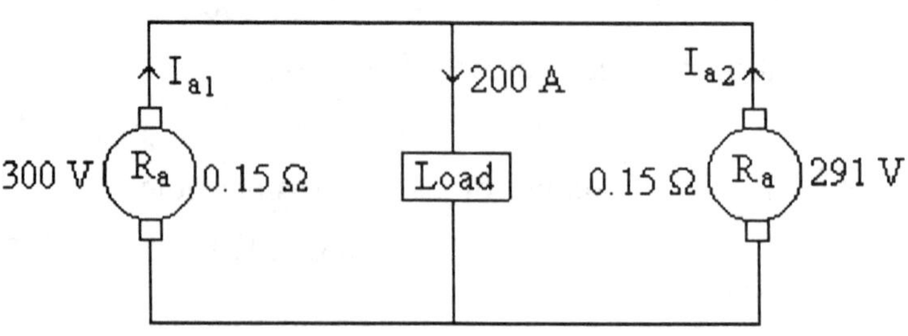

Fig. 6.10-0. Two generators connected in parallel

(contd)

<u>Refer to Fig. 6.10-0</u>

When the two generators are connected in parallel, the circuit is as shown in Fig. 6.10-0, with the field current neglected.

By application of Kirchhoff's second law, we get

$$300 - 291 = 0.15 (I_{a1} - I_{a2}) \qquad \text{(Eq. 6.10-1)}$$

By Kirchhoff's second law, we get

$$I_{a1} + I_{a2} = 200 \qquad \text{(Eq. 6.10-2)}$$

From Eq. 6.10-2,

$$I_{a1} = 200 - I_{a2}$$

Substituting this in Eq. 6.10-1, we get

$$300 - 291 = 0.15 (200 - I_{a2} - I_{a2})$$
$$9 = 0.15 (200 - 2 I_{a2})$$
$$9/0.15 = 200 - 2 I_{a2}$$
$$\therefore \quad I_{a2} = (200 - 9/0.15)/2$$
$$= (200 - 60)/2 = \underline{\textbf{70 A}}$$
$$\therefore \quad I_{a1} = 200 - 70 = \underline{\textbf{130 A}}$$

Terminal voltage, $V_T = E - I_{a1} R_a$
$$= 300 - (130 \times 0.15) = \underline{\textbf{280.5 V}}$$

Assumptions:

    (i)   The field current was negligible and did not affect the generated e.m.f.

    (ii)   Armature reaction was negligible.

    (iii)   Brush voltage drop was negligible.

1.   A four-pole, wave-wound armature has 450 conductors and is rotated
        at 1000 rev/min. Calculate the generated e.m.f if the useful flux per pole
        is 30 mWb.   *[Ans. 450 V]*

2.   A six-pole, lap-wound armature is rotated at 500 rev/min. If the useful flux
        per pole is 50 mWb, calculate the number of conductors if the generated
        e.m.f is 450 V.   *[Ans. 1080]*

3.   A four-pole, wave-wound armature having 57 slots with 12 conductors per
        slot is rotated under the influence of 0.02 W of useful flux per pole.
        Calculate the speed at which the armature is rotated to generate 250 V.
                                                                     [Ans. 548 r.p.m]

4.   An eight-pole d.c. generator has an armature wound with 702 conductors.
        Each conductor has a capacity to carry 40 A without overheating. The
        e.m.f generated in each conductor is 1.5 V. Calculate the e.m.f generated
        and the maximum armature current obtainable without overheating if the
        armature is connected (a) wave and (b) lap.
                                         [Ans. (a) 526.5 V, 80 A; (b) 132 V, 240 A]

5.   A four-pole, wave-wound armature has 151 turns and delivers 40 kW
        at 200 V. The armature has a resistance of 0.01 Ω and the useful flux
        per pole is 0.02 Wb. Taking the shunt field resistance as 50 Ω and
        neglecting voltage drop at the brushes, calculate the speed of the armature.
                                                                 [Ans. 1003 rev/min]

6.   A four-pole d.c. generator has a no-load voltage of 410 V when rotated at
        900 rev/min. Calculate the flux per pole if the armature is wave wound
        and has 39 slots with 16 conductors per slot.   *[Ans. 21.9 mWb]*

7.   A two-pole d.c. generator has a flux per pole in the air gap of 5 mWb.
        The rotating armature coil has 240 turns (480 conductors) connected
        in series. Calculate the average e.m.f at the brushes when the armature
        is rotated at 100 rev/min with negligible load current.   *[Ans. 40 V]*

8.   A shunt-connected d.c. generator has a field resistance of 120 Ω and
        feeds a 20 Ω resistor as load. The armature resistance is 2.5 Ω and the
        rotational losses in the generator are equivalent to an additional electrical
        shunt load of 100 Ω across the output terminals. Calculate the efficiency
        of the generator.   *[Ans. 73.3%]*

9.   A compound-wound d.c. armature has a full-load output of 120 A at 240 V. The shunt field resistance is 120 Ω and the effective armature resistance is 0.164 Ω. The sum of the iron, friction and windage losses is 1280 W at full load when the speed is 1000 rev/min. Calculate the torque required to turn the armature in (a) N m and (b) ft-lbf.

[Ans. (a) 315 N m, (b) 232 ft-lbf]

10.   Two d.c. shunt generators operate in parallel, and each has an armature resistance of 0.01 Ω and a field resistance of 110 Ω. Generator A has an e.m.f of 226 V and generator B an e.m.f of 224 V. Calculate the terminal voltage and power output of each generator when they are supplying a total load current of 996 A.

[Ans. 220 V, I_A = 600 A, I_B = 400 A, P_A = 131.56 kW, P_B = 87.56 kW]

CHAPTER   7

DIRECT CURRENT MOTORS

Introduction

There is no difference between the constructional features of a d.c. motor
and a d.c. generator. However, in terms of voltage, the generated e.m.f, E,
in a generator is greater than its terminal voltage V_T, whereas the terminal
voltage V of the motor (this is the supply voltage applied to the terminals
of the motor) is greater than the e.m.f, E, generated in the armature. The
generated e.m.f, E, in the armature of the motor is in opposition to the
supply voltage. Hence, it is sometimes referred to as the *back e.m.f.*

<u>Refer to Fig. 7.0-0</u>

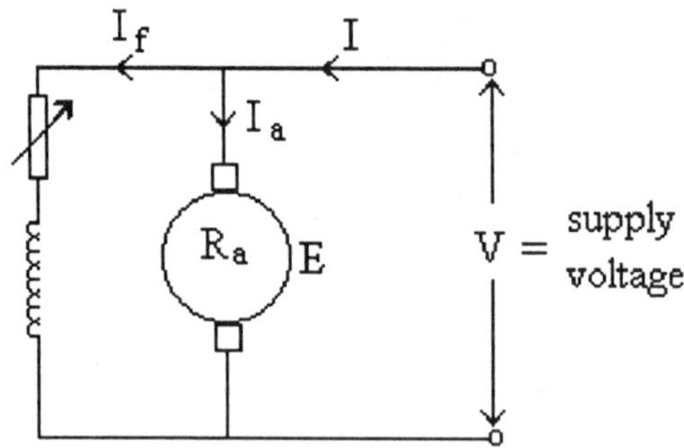

Fig. 7.0-0. Shunt-excited d.c. motor

The field current I_f is in the same direction whether the machine is a
generator or a motor. However, the armature current I_a is in the opposite
direction to that of a generator.

Let   E   = e.m.f generated by the armature

V   = supply voltage applied at the terminals

R_a = armature resistance

I_a = armature current

Then as a motor

$$E = V - I_a R_a \qquad \text{(Eq. 7.0-0)}$$

$$\text{or} \quad V = E + I_a R_a \qquad \text{(Eq. 7.0-1)}$$

D.c. motors and generators

Summary of Important Relationships

Legend:

$$E \quad = \text{generated e.m.f [V]}$$
$$V_T \quad = \text{(\textit{For a generator}) voltage obtained at the terminals [V]}$$
$$V \quad = \text{(\textit{For a motor}) voltage applied at the terminals [V]}$$
$$I_a \quad = \text{armature current [A]}$$
$$R_a \quad = \text{armature resistance } [\Omega]$$
$$R_f \quad = \text{armature field resistance } [\Omega]$$
$$I_f \quad = \text{shunt field current [A]}$$
$$\Phi \quad = \text{useful flux per pole} - \text{total flux entering armature [Wb]}$$
$$N \quad = \text{armature speed in rev/min}$$
$$T \quad = \text{torque acting upon armature [N m]}$$
$$p \quad = \text{number of pairs of poles}$$
$$c \quad = \text{number of parallel paths through the armature}$$
$$\quad = 2 \text{ for a wave-wound armature}$$
$$\quad = 2p \text{ for a lap-wound armature}$$
$$Z \quad = \text{number of armature conductors}$$
$$\quad = \text{number of slots x conductors per slot}$$

For a generator

$$E \quad > \quad V_T$$
$$\therefore \quad E \quad = V_T + I_a R_a \quad \text{volts}$$

For a motor

$$E \quad < \quad V$$
$$\therefore \quad E \quad = V - I_a R_a \quad \text{volts}$$

Motor speed – N

The e.m.f generated in the armature of the motor is given by the expression

$$E \quad = \frac{2 \, \Phi \, Z \, N \, p}{60 \, c} \qquad \text{(Refer to Eq. 6.0-0)}$$

For a given motor, the only variables are the speed N and the flux Φ. All other parameters (Z, p and c) are constant.

$$\therefore \quad E \quad \propto N \Phi \qquad \text{(Eq. 7.0-2)}$$
$$\text{and} \quad N \quad \propto E/\Phi \qquad \text{(Eq. 7.0-3)}$$

In practice, the speed of a motor may be controlled by varying either E or Φ or both.

Motor torque – T

The e.m.f equation is $E = \dfrac{2\,\Phi\,Z\,N\,p}{60\,c}$

From Eq. 7.0-1,

$$V = E + I_a R_a$$

The power supplied to the motor is given by the expression

$$V I_a = E I_a + I_a^2 R_a$$

where

$V I_a$ = power supplied to the armature

$I_a^2 R_a$ = power loss in the armature

$E I_a$ = mechanical power developed by the armature

In terms of torque, the mechanical power developed by the armature is

$$= (2\pi\,N\,T)/60 \quad\text{watts}$$

$$\therefore \quad (2\pi\,N\,T)/60 = E I_a$$

$$= \frac{2\,\Phi\,Z\,N\,p}{60\,c} \times I_a$$

and

$$T = \frac{2\,\Phi\,Z\,N\,p}{60\,c} \times I_a \times \frac{60}{2\pi\,N} \quad \text{N m}$$

For a given machine, Z, p and c are constant.

$$\therefore \quad T \propto \Phi \times I_a \qquad\qquad \text{(Eq. 7.0-4)}$$

For shunt motors

Assume unsaturated magnetic circuit.

$$\Phi \propto I_f$$

Then from Eq. 7.0-4

$$T \propto I_f \times I_a \qquad\qquad \text{(Eq. 7.0-5)}$$

From Eq. 7.0-0

$$E = V - I_a R_a$$

and

$$N \propto E/\Phi \qquad\qquad \text{(Refer to Eq. 7.0-3)}$$

$$\therefore \quad N \propto \frac{V - I_a R_a}{I_f} \qquad\qquad \text{(Eq. 7.0-6)}$$

For series motors

Assume unsaturated magnetic circuit.

$$\Phi \propto I_f$$

Then from Eq. 7.0-5

$$T \propto I_a^2 \qquad\qquad \text{(Eq. 7.0-7)}$$

Also

$$E \propto N I_a \qquad\qquad \text{(Refer to Eq. 7.0-2)}$$

and

$$N \propto E/I_a \qquad\qquad \text{(Eq. 7.0-8)}$$

Example 7.1

The armature circuit of a d.c. machine has a resistance of 0.2 Ω and is connected to a 240 V supply. Calculate the generated e.m.f when running

(a)   as a generator delivering 120 A

(b)   as a motor taking 80 A

Solution

(a)   As a d.c. generator

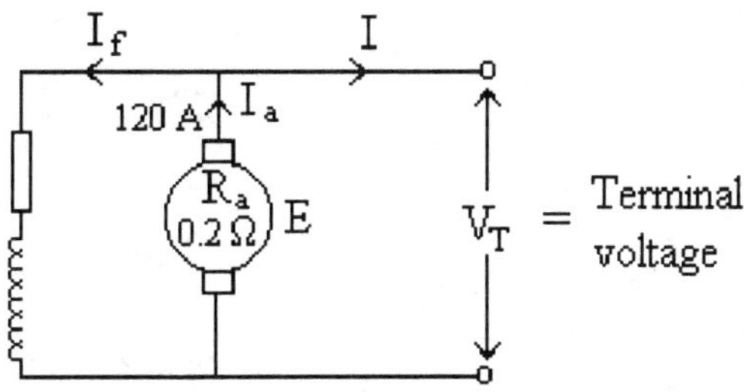

Fig. 7.1-0. D.C. Generator

<u>Refer to Fig. 7.1-0</u>

$$\text{Voltage drop in armature} = I_a R_a = 120 \times 0.2 = \underline{24 \text{ V}}$$

$$\text{Generated e.m.f, } E = V_T + I_a R_a$$

$$= 240 + 24 \qquad = \underline{\mathbf{264 \text{ V}}}$$

(b)   As a d.c. motor

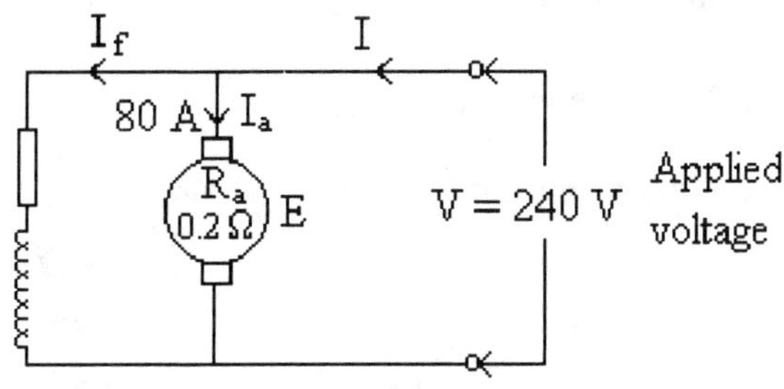

Fig. 7.1-1. D.C. Motor

<u>Refer to Fig. 7.1-1</u>

$$\text{Voltage drop in armature} = I_a R_a = 80 \times 0.2 = \underline{16 \text{ V}}$$

$$\text{Generated e.m.f, } E = V - I_a R_a$$

$$= 240 - 16 \qquad = \underline{\mathbf{224 \text{ V}}}$$

Example 7.2

A four pole d.c. motor is fed from a 500 V supply and takes an armature current of 50 A. The armature is wave-wound with 1200 conductors. If the useful flux per pole is 0.025 Wb and the armature circuit resistance is 0.4 Ω, calculate the speed at which the armature is rotating.

Solution

$$V \quad = E + I_a R_a \qquad \text{(Refer to Eq. 7.0-1)}$$
$$E \quad = V - I_a R_a$$
$$= 500 - (50 \times 0.4)$$
$$= 500 - 20 \qquad = \underline{480 \text{ V}}$$

Now
$$E \quad = \frac{2 \, \Phi \, Z \, N \, p}{60 \, c} \qquad \text{(Refer to Eq. 6.0-0)}$$

$$480 \quad = \frac{2 \times 0.025 \times 1200 \times N \times 2}{60 \times 2}$$

$\therefore$
$$N \quad = {}^{480}/(0.025 \times 20 \times 2)$$

$$= \underline{\mathbf{480 \text{ rev/min}}}$$

Example 7.3

A six-pole d.c. motor takes an armature current of 120 A from an applied voltage of 480 V. The armature is lap-wound with 860 conductors and has a resistance of 0.2 Ω. The useful flux per pole is 0.05 Wb. Calculate (a) the speed and (b) the gross torque developed by the armature.

Solution

(a)   Generated e.m.f, E
$$= V - I_a R_a \qquad \text{(Refer to Eq. 7.0-0)}$$
$$= 480 - (120 \times 0.2)$$
$$= 480 - 24 \qquad = \underline{456 \text{ V}}$$

$$E \quad = \frac{2 \, \Phi \, Z \, N \, p}{60 \, c}$$

$\therefore$
$$N \quad = \frac{E \times 60 \, c}{2 \, \Phi \, Z \, p}$$

$$= \frac{456 \times 60 \times 2 \times 3}{2 \times 0.05 \times 860 \times 3}$$

$$= \frac{456 \times 6}{0.05 \times 86} \qquad = \underline{\mathbf{636 \text{ rev/min}}}$$

(contd)

(b)   Mechanical power developed by armature is $= E\,I_a$ watts.

Mechanical power developed $= (2\pi\,N\,T)/60$

Hence $(2\pi\,N\,T)/60 = E\,I_a$

$\therefore$ $T = \dfrac{60 \times E \times I_a}{2\pi\,N}$

$= \dfrac{60 \times 456 \times 120}{2 \times 3.142 \times 636}$ $= \textbf{821.5 N m}$

Example 7.4

A d.c. machine has the following details:

 Poles, 4

 Flux per pole, 59 mWb

 Armature winding, wave-wound

 Number of slots, 48

 Conductors per slot, 4

 Armature resistance, 0.2 Ω

When the machine was connected to a 400 V supply, it ran as a motor at a speed of 1050 rev/min.

Calculate the back e.m.f; hence find the value of the gross torque developed by the motor in newton meters and in ft-lb.

Solution

Back e.m.f, $E = \dfrac{2\,\Phi\,Z\,N\,p}{60\,c}$

$= \dfrac{2 \times 59 \times 10^{-3} \times 48 \times 4 \times 1050 \times 2}{60 \times 2}$

$= 2 \times 59 \times 10^{-3} \times 4 \times 4 \times 210$

$= 118 \times 10^{-3} \times 16 \times 210$

$= 118 \times 10^{-3} \times 3360$

$= 396480 \times 10^{-3}$ $= \underline{396.48\ V}$

(contd)

The mechanical power developed by the armature is

$$= E I_a$$

Gross torque, $(2\pi N T)/60 = E I_a$

$$\therefore \quad I_a = (V - E)/R_a$$

$$= (400 - 396.48)/0.2 = \underline{17.6\ A}$$

and

$$T = \frac{E I_a \times 60}{2\pi N}$$

$$= \frac{396.48 \times 17.6 \times 60}{2 \times 3.142 \times 1050} = \mathbf{63.45\ N\ m}$$

Torque, T (N m)

$$= \frac{63.45 \times 0.225}{0.3048} = \mathbf{46.84\ ft\text{-}lb}$$

Example 7.5

Briefly explain why the current supplied to a d.c. motor increases as the motor is loaded. A four-pole d.c. shunt motor takes an armature current of 25 A from a 400 V supply. The armature is wave-wound and has 59 slots, 8 conductors per slot and a resistance of 0.2 Ω. Calculate the useful flux per pole when the speed is 1050 rev/min.

Solution

From the equation $E = (2\,\Phi\,Z\,N\,p)/60\,c$, the (*back e.m.f*) E is proportional to the speed N. Thus when the motor is loaded, the speed falls, which is accompanied by a fall in E, and as a result the current taken from the supply is increased. The increase in current will cause a further decrease in E ($E = V - I_a R_a$). However, since this reduction is almost negligible, the speed N will automatically adjust until the input power balances the output power plus the losses.

$$E = \frac{2\,\Phi\,Z\,N\,p}{60\,c}$$

$$\Phi = \frac{E \times 60\,c}{2\,Z\,N\,p}$$

$$E = V - I_a R_a = 400 - (25 \times 0.2) = \underline{395\ V}$$

$$\therefore \quad \Phi = \frac{395 \times 60 \times 2}{2 \times 59 \times 8 \times 1050 \times 2} = \frac{\mathbf{0.0239\ Wb}}{(\mathbf{23.9\ mWb})}$$

Example 7.6

A 120 V, 2 h.p d.c. motor has an armature resistance of 0.2 Ω.
Determine the armature current

    (a)   when the back e.m.f, E, is 115 V

    (b)   at the instant the machine is connected to the supply; state the necessary precautions, if any, that should be taken

Solution

(a)   From the expression

$$E = V - I_a R_a$$

$$I_a = \frac{V - E}{R_a}$$

$$= \frac{120 - 115}{0.2} \qquad = \underline{\mathbf{25\ A}}$$

(b)   At the instant of starting, $E = 0$.

$$I_a = \frac{120 - 0}{0.2} \qquad = \underline{\mathbf{600\ A}}$$

This current is much too high for a two h.p motor, and without protection, the armature will obviously fry. The normal means of protection is by either fuses or a circuit breaker. It is essential to limit the armature current at starting to a safe value. In practice, additional resistance is added in series with the armature circuit to limit the current to a nominal value during starting. This resistance is cut out in steps as the motor picks up speed.

Example 7.7

A 240 V d.c. motor has an armature resistance of 0.2 Ω and takes 25 A when running at 1500 rev/min. Calculate at what speed the motor would run if the armature current was 40 A and a 0.2-Ω resistor is connected in series with the armature.

Solution

$$N = 1500\ rev/min$$

$$E = V - I_a R_a$$
$$= 240 - (25 \times 0.2) \qquad = \underline{235\ V}$$

(contd)

New value of N

$$I_a = 40 \text{ A}$$

Additional resistance in series with armature $R_{se} = 0.2\ \Omega$

$$E = V - I_a (R_a + R_{se})$$
$$= 240 - 40 (0.2 + 0.2)$$
$$= \underline{224 \text{ V}}$$

Now $\quad E \propto k\,N\,\Phi \qquad$ (Refer to Eq. 7.0-2)

Since $k\,\Phi$ is constant, $E \propto N$

$$\text{Let} \quad N_1 = 1500 \text{ rev/min}$$
$$N_2 = \text{new speed}$$
$$E_1 = 235 \text{ V}$$
$$E_2 = 224 \text{ V}$$

Thus $\qquad E_1/E_2 = N_1/N_2$

and $\qquad N_2 = N_1 \times (E_2/E_1)$

$$= 1500 \times (^{224}/235) \quad = \underline{\mathbf{1430\ rev/min}}$$

Example 7.8

A series motor having an armature resistance of 0.2 Ω and a field resistance of 1.3 Ω takes a current of 25 A from a 240 V supply when running at 1000 rev/min. Calculate the new speed the motor will run at when a 3.5 Ω resistor is connected in series with the motor, the load being modified so that the current taken remains at 25 A.

Solution

For a series motor, $N \propto E/I_a \qquad$ (Refer to Eq. 7.0-8)

$$\text{Let} \quad N_1 = 1000 \text{ rev/min (old speed)}$$

$$N_2 = \text{new speed with additional resistance}$$

$$\therefore \quad N_1 \propto \frac{V - I_a (R_a + R_f)}{I_a}$$

and $\qquad N_2 \propto \dfrac{V - I_a (R_a + R_f + R_{se})}{I_a}$

(contd)

$$\therefore \quad N_1/N_2 = \frac{[V - I_a(0.2 + 1.3)]/I_a}{[V - I_a(0.2 + 1.3 + 3.5)]/I_a}$$

$$= \frac{[V - I_a(1.5)]/I_a}{[V - I_a(5)]/I_a}$$

$$\therefore \quad N_2 = N_1 \frac{[V - 5 I_a]}{V - 1.5 I_a}$$

$$= 1000 \frac{[240 - (5 \times 25)]}{240 - (1.5 \times 25)}$$

$$= 1000 [^{115}/202.5]$$

$$= \underline{\mathbf{568 \ rev/min}}$$

Example 7.9

A 220 V shunt motor has an armature resistance of 0.5 Ω and takes a current of 20 A when running at 750 rev/min. Calculate the resistance to be placed in series with the armature to reduce the speed to 500 rev/min. Determine the new speed if, when the resistance is inserted, the load is altered to reduce the armature current to 14 A. Neglect the effect of armature reaction.

Solution

It is assumed that the field current remains constant since no mention was made of it.

$$N \quad \propto \quad E/\Phi \qquad \text{(Refer to Eq. 7.0-3)}$$

Assume an unsaturated magnetic circuit, $\Phi \propto I_f$. Therefore

$$N \quad \propto \quad E/I_f$$

and

$$E \quad \propto \quad I_f N$$

Let $\quad E_1 = 220$ V

$\quad E_2 = $ new e.m.f

$\quad N_1 = 750$ rev/min

$\quad N_2 = 500$ rev/min

Then $\quad E_1/E_2 = N_1/N_2$

and $\quad E_2 = E_1 \times N_2/N_1$

$$E_1 = V - I_a R_a$$
$$= 220 - (20 \times 0.5) \quad = \underline{210 \text{ V}}$$

$$\therefore \quad E_2 = 210 \times (^{500}/750) \quad = \underline{140 \text{ rev/min}}$$

Now
$$E_2 = V - I_a (R_a + R_{se})$$
$$140 = 220 - 20 (0.5 + R_{se})$$
$$= 220 - 10 - 20 R_{se}$$
$$20 R_{se} = 220 - 10 - 140$$
$$= 220 - 150$$
$$= 70$$

$$\therefore \quad R_{se} = 70/20 \qquad = \underline{\mathbf{3.5 \ \Omega}}$$

Final speed $I_a = 14$ A

Let E_3 = new e.m.f at this speed

Then
$$E_3 = V - I_a (R_a + R_{se})$$
$$= 220 - 14 (0.5 + 3.5)$$
$$= 220 - 56 \qquad = \underline{164 \text{ V}}$$

Let N_3 = new speed

Then
$$E_1/E_3 = N_1/N_3$$

and
$$N_3 = N_1 \times {}^{E_3}/E_1$$
$$= 750 \, (^{164}/210) \qquad = \underline{\mathbf{586 \text{ rev/min}}}$$

Example 7.10

A d.c. series motor takes 50 A and runs at 500 rev/min when connected to a 220 V supply. Calculate the value of the resistance which, when inserted in series with the armature, will reduce the speed of the motor to 300 rev/min, the load torque then being one-half of its previous value. The resistance of the motor is 0.1 Ω (i.e., $R_a + R_f$), and the flux is proportional to the field current.

Solution

$$\text{Flux} \quad \Phi \quad \propto \quad I_f$$

$$\text{Torque} \quad T \quad \propto \quad I_a^2 \qquad \qquad \text{(Refer to Eq. 7.0-7)}$$

(contd)

Let T_1 = unity

Then T_2 = $1/2$

I_{a1} = 50 A

Now $\qquad T_1/T_2 = I_{a1}^2/I_{a2}^2$

$\qquad 1/\tfrac{1}{2} = 50^2/I_{a2}^2$

$\qquad 2 = 50^2/I_{a2}^2$

$\qquad I_{a2}^2 = 50^2/2$

$\therefore \qquad I_{a2} = \sqrt{(50^2/2)} \quad = 50/\sqrt{2}$

Also $\qquad N \propto E/I_a \qquad$ (Refer to eq. 7.0-8)

$\qquad\qquad \propto [V - I_a (R_a + R_f)]/I_a$

$\therefore \qquad N_1/N_2 = E_1/E_2$

$$= \frac{[V - I_{a1}(R_a + R_f)]/I_{a1}}{[V - I_{a2}(R_a + R_f + R_{se})]/I_{a2}}$$

where N_1 = 500 rev/min

$\qquad N_2$ = 300 rev/min

$\qquad V$ = 220 V

$\qquad R_{se}$ = inserted series resistance

$\qquad I_{a1}$ = 50 A

$\qquad I_{a2}$ = $50/\sqrt{2}$

$$500/300 = \frac{[220 - (50 \times 0.1)]/50}{[220 - \{(50/\sqrt{2})(0.1 + R_{se})\}]/(50/\sqrt{2})}$$

$$5/3 = \frac{220 - 5}{220 - 35.36(0.1 + R_{se})} \times \frac{50}{\sqrt{2} \times 50}$$

$$= \frac{215}{220 - 3.536 - 35.36 R_{se}} \times \frac{1}{\sqrt{2}}$$

$$\frac{215}{\sqrt{2}} = (5/3)(220 - 3.536 - 35.36 R_{se})$$

$$\frac{3}{5} \times \frac{215}{\sqrt{2}} = 216.464 - 35.36 R_{se}$$

$$R_{se} = [216.464 - (3/5 \times 215/\sqrt{2})]/35.36$$

$$= [216.464 - 91.2]/35.36 \quad = \mathbf{3.54\ \Omega}$$

Example 7.11

A 220 V d.c. shunt motor takes 25 A at a speed of 1000 rev/min. The armature resistance is 0.1 Ω and the field resistance is 100 Ω. Calculate the speed at which the motor will run if the load torque is halved, a resistance of 50 Ω is added onto the field circuit and a 5 Ω resistance is inserted in the armature circuit. Assume field flux to be proportional to field current.

Solution

Before insertion of resistances in the field and armature circuits

$$V = 220 \text{ V} \qquad\qquad R_a = 0.1\ \Omega$$
$$R_{f1} = 100\ \Omega \qquad\qquad I = 25 \text{ A}$$
$$I_{f1} = V/R_{f1} \qquad\qquad N_1 = 1000 \text{ rev/min}$$
$$= 220/100 \qquad\qquad I_{a1} = I - I_{f1}$$
$$= 2.2 \text{ A} \qquad\qquad = 25 - 2.2 = 22.8 \text{ A}$$
$$E_1 = V - I_{a1}\,R_{a1}$$
$$= 220 - (22.8 \times 0.1) = 217.72$$

After insertion of resistances in the field and armature circuits

$$V = 220 \text{ V} \qquad\qquad R_{a2} = 0.1 + 5 = 5.1\ \Omega$$
$$R_{f2} = 100 + 50 \qquad\qquad I_{f2} = V/R_{f2}$$
$$= 150\ \Omega \qquad\qquad = 220/150 = 1.467 \text{ A}$$

$$\text{Let} \quad T_1 = \text{unity}$$
$$\therefore \quad T_2 = {}^1\!/_2$$
$$\text{and} \quad T_1/T_2 = 2$$

Since the torque was halved, we need to find the value of I_{a2}.

$$\text{Thus} \quad T_1/T_2 = (I_{a1}\,I_{f1})/(I_{a2}\,I_{f2}) \qquad (\text{Note: } T \propto I_a\,I_f)$$
$$2\,(I_{a2}\,I_{f2}) = I_{a1}\,I_{f1}$$
$$I_{a2} = (I_{a1}\,I_{f1})/2\,I_{f2}$$
$$= (22.8 \times 2.2)/(2 \times 1.467) = \underline{17.1 \text{ A}}$$

(contd)

$$N \propto (V - I_a R_a)/I_f \qquad \text{(Refer to Eq. 7.0-6)}$$

$$N_1/N_2 = \frac{(V - I_{a1} R_{a1})/I_{f1}}{(V - I_{a2} R_{a2})/I_{f2}}$$

$$N_2 = N_1 \frac{(V - I_{a2} R_{a2})/I_{f2}}{(V - I_{a1} R_{a1})/I_{f1}}$$

$$= N_1 \frac{(V - I_{a2} R_{a2}) \times I_{f1}}{(V - I_{a1} R_{a1}) \times I_{f2}}$$

$$= 1000 \frac{[220 - (17.1 \times 5.1)]}{[220 - (22.8 \times 0.1)]} \times \frac{2.2}{1.467}$$

$$= 1000 \frac{[220 - 87.21]}{[220 - 2.28]} \times 1.5$$

$$= 1000 \times \left[\frac{132.79}{217.72}\right] \times 1.5$$

$$= 1000 \times 0.6099 \times 1.5 \qquad = \underline{\textbf{915 rev/min}}$$

Example 7.12

A 240 V shunt motor has a full-load speed of 750 rev/min and a full-load armature current of 20 A. The armature resistance is 1 Ω. The maximum current during the starting period is 30 A. Calculate (a) the total resistance of the starter, (b) the resistance between the first two studs if the starter handle is to be moved to the second stud when the motor speed is 250 rev/min and (c) the current immediately before the starter handle is moved to the second stud.

Solution

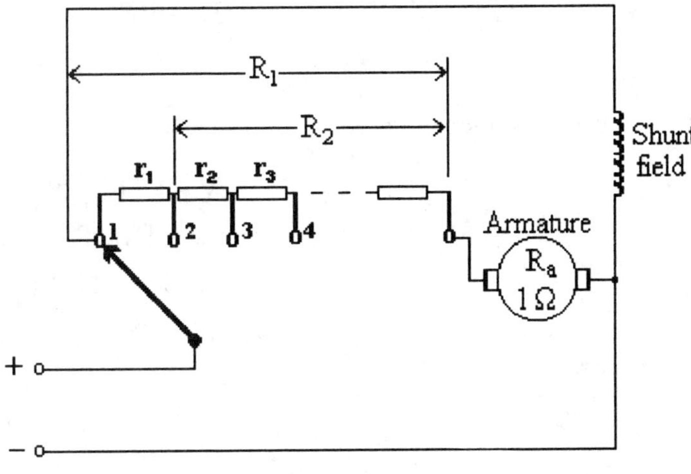

Fig. 7.12-0

(contd)

(a)   <u>Refer to Fig. 7.12-0</u>

At start, the e.m.f $E = 0$

Armature circuit resistance $= R_a + R_1$

$$= {}^{240}/30 = \underline{8\ \Omega}$$

$$R_1 = 8 - 1 \qquad\qquad = \underline{7\ \Omega}$$

At 750 rev/min (N_1)

$$E_1 = V - I_a R_a$$

$$= 240 - (20 \times 1) \qquad = \underline{220\ V}$$

At 250 rev/min (N_2)

$$E_1/E_2 = N_1/N_2$$

$$E_2 = E_1 \times {}^{N_2}/N_1$$

$$= 220 \times {}^{250}/750 \qquad = \underline{73.33\ V}$$

Armature circuit resistance on second stud is

$$= R_a + R_2$$

Now $\qquad E_2 = V - I_a (R_a + R_2)$

$$R_a + R_2 = (V - E_2)/I_a$$

$$= (240 - 73.33)/30 \quad = \underline{5.557\ \Omega}$$

$\therefore \qquad R_2 = 5.557 - R_a$

$$= 5.557 - 1 \qquad = \underline{4.557\ \Omega}$$

(b)   Resistance between the first two studs is r_1

and $\qquad r_1 = R_1 - R_2$

$$= 7 - 4.557 \qquad = \underline{\mathbf{2.443\ \Omega}}$$

(c)   Current immediately before moving to the second stud

$$E_2 = V - I_a (R_a + R_1)$$

$$I_a = (V - E_2)/(R_a + R_1)$$

$$= (240 - 73.33)/(1 + 7)$$

$$= {}^{166.67}/8 \qquad = \underline{\mathbf{20.83\ A}}$$

Example 7.13

A 200 V shunt motor takes a current of 26 A at full load when running at 500 rev/min. The resistances of the armature and field are 0.2 Ω and 200 Ω, respectively. To control the speed of the motor, a 1.8 Ω resistor is connected in series with the armature. Calculate

(a) *the speed at which the motor will run when supplying full-load torque*

(b) *the torque at which the motor will stall with the 1.8 Ω resistor in the circuit in terms of the full-load torque.*

Solution

(a) *Without series resistor – R_{se} (1.8 Ω)*

$$\text{Field current,} \quad I_f = {}^V\!/R_f = 200/200 = \underline{1.0 \text{ A}}$$

$$\text{Armature current,} \quad I_{a1} = 26 - 1.0 = \underline{25 \text{ A}}$$

$$\therefore \quad E_1 = V - I_a R_a$$
$$= 200 - (25 \times 0.2) = \underline{195 \text{ V}}$$

With series resistance – R_{se} (1.8 Ω)

Torque T and I_f have to remain constant (with I_f constant, Φ is constant).

Therefore, I_a remains unchanged.

$$\therefore \quad E_2 = V - I_a (R_a + R_{se})$$
$$= 200 - 25 (0.2 + 1.8)$$
$$= 200 - 50 = 150 \text{ V}$$

$$N \propto {}^E\!/\Phi \qquad \text{(Refer to Eq. 7.0-3)}$$

$$\text{or} \quad E \propto N \Phi$$

$$\therefore \quad E_1/E_2 = {}^{N_1}\!/N_2$$

$$N_2 = N_1 \times {}^{E_2}\!/E_1$$

$$= 500 \times {}^{150}\!/195 = \underline{\textbf{385 rev/min}}$$

(contd)

(b)   *Stalling torque*

When the motor is stalled

$$E = 0$$

$$E = V - I_{a2}(R_a + R_{se})$$

$$\therefore \quad I_{a2} = V/(R_a + R_{se})$$

$$= 200/(0.2 + 1.8) \quad = \underline{100\ A}$$

Now $\qquad T \propto \Phi I_a \qquad$ (Refer to Eq. 7.0-4)

$$\therefore \quad T_1/T_2 = I_{a1}/I_{a2}$$

Stalling torque, $\quad T_2 = T_1 \times I_{a2}/I_{a1}$

$$= T_1 \times 100/25$$

$$= 4\ T_1$$

That is, the stalling torque is four times the full-load torque.

Example 7.14

A 400 V d.c. series motor has a total armature and field resistance of 0.2 Ω. A variable resistance R is connected in series with the armature for speed adjustment. With a given load and R = 0, the current is 25 A and the speed is 1000 rev/min. On another load with R = 2 Ω, the current is 20 A. Calculate the new speed and the ratio of power outputs of the motor in the two cases. Assume that the field flux is proportional to the current and ignore mechanical and iron losses in the motor.

Solution

Case 1, R = 0

$$E_1 = V - I_{a1}R_a$$

$$= 400 - (25 \times 0.2) \quad = \underline{395\ V}$$

Case 2, R = 2 Ω

$$E_2 = V - I_{a2}(R_a + R)$$

$$= 400 - 20(0.2 + 2)$$

$$= 400 - 44 \quad = \underline{356\ V}$$

(contd)

$$\text{Now} \qquad E \quad \propto N\, I_a \qquad\qquad \text{(Refer to Eq. 7.0-2)}$$

$$\therefore \qquad E_1/E_2 \quad = N_1\, I_{a1}/N_2\, I_{a2}$$

$$\text{and} \qquad N_2 \quad = N_1\,(E_2/E_1 \times I_{a1}/I_{a2})$$

$$= 1000\,(^{356}/395 \times {}^{25}/20)$$

$$= 1000\,(0.901 \times 1.25) = \underline{1126 \text{ rev/min}}$$

$$\text{Now} \qquad T \quad \propto I_a^2 \qquad\qquad \text{(Refer to Eq. 7.0-7)}$$

$$P \quad = (2\pi N\, T)/60$$

$$\therefore \qquad P \quad \propto \omega\, T$$

$$\text{Substituting for } T, \qquad P \quad \propto \omega\, I_a^2$$

$$T \quad \propto \Phi \times I_a \qquad\qquad \text{(Refer to Eq. 7.0-4)}$$

$$\text{and} \qquad \Phi \quad \propto I_a$$

$$\therefore \qquad T \quad \propto I_a^2$$

Therefore ratio of output powers

$$P_1/P_2 \quad = \omega\, I_{a1}^2/\omega\, I_{a2}^2$$

$$= \frac{(2\pi\, N_1\, T)/60}{(2\pi\, N_2\, T)/60} \times (I_{a1}/I_{a2})^2$$

$$= (N_1/N_2) \times (I_{a1}/I_{a2})^2$$

$$= {}^{1000}/1126 \times (^{25}/20)^2$$

$$= 0.888 \times (1.25)^2 \quad = \underline{\mathbf{1.388}}$$

Example 7.15

A 500 V, 50 h.p series motor runs at 750 rev/min on full-load.
If the load torque is adjusted to 200 ft-lb and a 5 Ω resistor is included
in series with the armature, calculate the speed at which the motor will
run. Neglect all losses and assume the magnetic circuit to be unsaturated.

Solution

All losses are neglected.

$\therefore$ Armature resistance R_a is ignored

and   motor output assumed $= 50$ h.p

(contd)

$$\text{Full-load output power} = 50 \times 33000 \text{ ft-lb/min}$$

$$\text{Now} \quad 2\pi \, N \, T = 50 \times 33000$$

$$\therefore \quad \text{Torque,} \quad T = \frac{50 \times 33000}{2\pi \, N}$$

$$= \frac{50 \times 33000}{2 \times 3.142 \times 750} = \underline{350.1 \text{ ft-lb}}$$

$$T \propto I_a^2$$

$$\text{or} \quad I_a \propto \sqrt{T}$$

$$\text{On full-load,} \quad I_a = \frac{\text{h.p} \times 746}{V}$$

$$= \frac{50 \times 746}{500} = \underline{74.6 \text{ A}}$$

$$\text{Now} \quad (I_{a1}/I_{a2})^2 = T_1/T_2$$

$$\text{and} \quad I_{a2} = I_{a1} \times (T_2/T_1)^{1/2}$$

$$\text{where} \quad T_1 = 350.1 \text{ ft-lb}$$

$$T_2 = 200 \text{ ft-lb}$$

$$I_{a1} = 74.6 \text{ A}$$

$$\text{Therefore,} \quad I_{a2} = 74.6 \times (200/350.1)^{1/2}$$

$$= 74.6 \times 0.7558 = \underline{56.39 \text{ A}}$$

With 5 Ω resistor in series with the armature

$$E = V - I_{a2} R_a$$

$$= 500 - (56.39 \times 5) = \underline{218.05 \text{ V}}$$

$$E = \frac{2 \, \Phi \, Z \, N \, p}{60 \, c}$$

where Φ, Z, p and c are the usual armature constants

$$\therefore \quad [E_1/E_2] = (N_1 \, \Phi_1)/(N_2 \, \Phi_2) = (N_1 \, I_{a1})/(N_2 \, I_{a2})$$

$$\text{That is,} \quad \frac{500}{218.05} = \frac{750}{N_2} \times \frac{74.6}{56.39}$$

$$N_2 = \frac{218.05 \, (750 \times 74.6)}{500 \times 56.39}$$

$$= \frac{218.05 \times 55950}{500 \times 56.39} = \textbf{433 rev/min}$$

Example 7.16

A 240 V d.c. motor drives a water pump which delivers 800 gal of water per minute against a head of 60 feet. The efficiency of the motor is 80% and that of the pump is 60%. Calculate the horsepower of the motor and the current taken from the supply.

(Take 1 gallon of water = 10 lb)

Solution

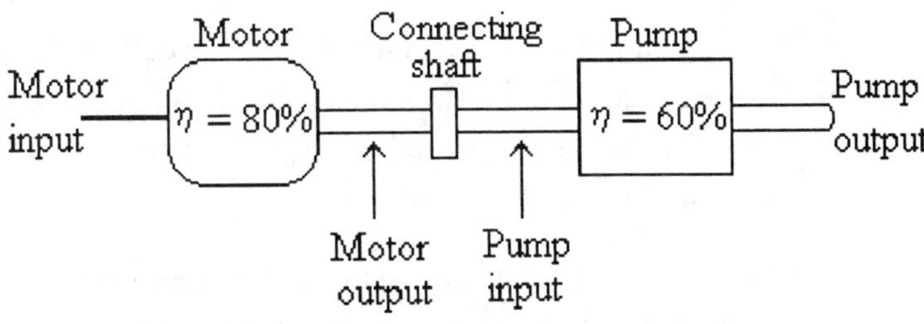

Fig. 7.16-0

<u>Refer to Fig. 7.16-0</u>

Work done by pump per minute

$$= 800 \times 10 \times 60 \qquad = \underline{480000 \text{ ft-lb}}$$

$$1 \text{ h.p} \qquad = 33000 \text{ ft-lb}$$

$$\text{Output horsepower of pump} = {}^{480000}/_{33000} \qquad = \underline{14.545 \text{ h.p}}$$

$$\text{Input power to pump} = {}^{\text{output}}/_{\text{efficiency}}$$

$$= {}^{14.545}/_{0.60} \qquad = \underline{24.242 \text{ h.p}}$$

$$\text{Output power of motor} = \text{input power to pump}$$

$$= \underline{\mathbf{24.242 \text{ h.p}}}$$

$$\text{Input power of motor} = {}^{\text{output}}/_{\text{efficiency}}$$

$$= {}^{24.242 \times 746}/_{0.8} \qquad = \underline{22605.665 \text{ W}}$$

$$\text{Current taken by motor} = {}^{\text{power}}/_{\text{voltage}}$$

$$= {}^{22605.665}/_{240} \qquad = \underline{\mathbf{94.2 \text{ A}}}$$

1. Derive the e.m.f equation for a d.c. machine.
    A d.c. shunt motor has an armature resistance of 0.5 Ω and is connected to a 200 V supply. If the armature current taken by the motor is 20 A, determine the e.m.f generated by the armature.   [*Ans. 190 V*]

2. A 230 V d.c. series motor has an armature resistance of 0.3 Ω and a field resistance of 0.4 Ω . The speed is 700 rev/min when the current is 25 A. Assuming the flux to be proportional to current, determine the speed of the motor when it takes 35 A from the supply.

    [*Ans. 484 rev/min*]

3. A 200 V d.c. series motor has an armature resistance of 0.05 Ω and a field resistance of 0.15 Ω. The machine takes 50 A when running at 1200 rev/min. Calculate the speed when the current is 75 A if the flux is directly proportional to the field current.   [*Ans. 778 rev/min*]

4. A 110 V d.c. motor has an armature resistance of 0.1 Ω and runs at 1000 rev/min when taking an armature current of 50 A. Determine the speed at which the motor will run if a resistor of 0.05 Ω is placed in series with the armature.   [*Ans. 976 rev/min*]

5. The armature resistance of a 200 V shunt motor is 0.4 Ω and the no-load armature current is 2 A. When loaded and taking an armature current of 50 A, the motor runs at 1200 rev/min. Calculate the approximate no-load speed and state any assumption made in the calculation.

    [*Ans. 1328 rev/min*]

6. A four-pole motor has a wave-wound armature with 594 conductors. The armature current is 40 A and the useful flux per pole is 7.5 mWb. Calculate the total horsepower developed when running at 1380 rev/min. Why does this calculated horsepower differ from the brake horsepower?
    [*Ans. 10.98 h.p*]

7. A shunt machine has an armature resistance of 0.05 Ω and a field resistance of 50 Ω. When connected across a 200 V supply, it runs as a motor at 500 rev/min and takes a total input of 100 kW. Determine the speed if the machine is to be driven as a generator to supply a load of 100 kW at a terminal voltage of 200 V.   [*Ans. 642 rev/min*]

8.	Explain why a direct current motor is not connected directly to the mains
when starting.
A 240 V d.c. motor has an armature resistance of 0.6 Ω. How would
you prevent the starting current from exceeding the safe value of 30 A?
A d.c. shunt motor which has an armature resistance of 0.15 Ω is to be
operated from a 200 V d.c. supply. The starting current is to be limited to
40 A. Calculate the resistance required when starting from rest.

[*Ans. 4.85 Ω*]

9.	A shunt motor takes 25 A on full-load from a 100 V d.c. supply. The
armature and field winding resistances are 0.3 Ω and 100 Ω, respectively.
The losses incurred in friction, windage and the iron circuit amount to
300 W. Determine the brake horsepower and the efficiency of the motor on
full-load.	[*Ans. 2.58 h.p, 77.09%*]

10.	A series motor developed 5 h.p when taking 50 A at a speed of
750 rev/min. Calculate the starting torque in ft-lb if the starting current is
limited to 75 A. Neglect armature reaction and assume that the flux is
proportional to the current.	[*Ans. 78.75 ft-lb*]

11.	A 440 V d.c. series motor has an armature and field coil resistance of
0.3 Ω. The motor has a regulating resistance of R Ω for speed adjustment.
On a certain load with R zero, the current is 20 A and the speed is
1200 rev/min. With another load and R = 3 Ω, the current is 15 A.
Determine the new speed and also the ratio of the two values of power
output of the motor. Assume that the field strength at 15 A is 80% of
that at 20 A.	[*Ans. 1350 rev/min, 1.48*]

12.	Enumerate the losses which occur in d.c. machines and state how each
loss varies with load.
A 300 V d.c. shunt motor takes a current of 3 A at no-load. Calculate the
output horsepower and efficiency of the motor when the full-load current
is 35 A. The armature and field resistances are 0.3 Ω and 460 Ω, respectively. Determine also the efficiency at half-load current.

[*Ans. 12.4 h.p, 88.1%, 81.27%*]

13.	A 250 V, four-pole series motor has a wave-wound armature with 788
conductors. The motor develops 10 shaft horsepower when taking 35 A.
The useful flux per pole is 25 mWb and the motor resistance is 1 Ω.
Calculate (a) the total torque and (b) the output torque in N m and in ft-lb.
[*Ans. (a) 219 N m, 161.66 ft-lb (b) 217.88 N m, 160.84 ft-lb*]

14. A 240 V d.c series motor takes a current of 10 A when running at
1000 rev/min. The armature resistance is 0.8 Ω and the field resistance is
0.2 Ω. Calculate the motor current when a 0.2 Ω resistor is connected
in parallel with the field and the speed is adjusted to 600 rev/min.
Determine also the ratio between the total torques developed under each of
the two above conditions. Assume that the flux is proportional to the field
current.

[Ans. 30.8 A, $T_1/T_2 = 0.211$ (i.e, $T_1 = 0.211\,T_2$ or $T_2 = 4.74\,T_1$)]

15. A 10 h.p, 230 V motor with an armature resistance of 0.2 Ω has
an efficiency of 85% at full-load. If the maximum permissible current at
starting is the full-load current and if the current is allowed to fall to 75%
of its full-load value before the resistance is altered, determine the
resistance necessary in the first three steps of the starter.

[Ans. $r_1 = 5.827\ \Omega$; $r_2 = 4.32\ \Omega$; $r_3 = 3.19\ \Omega$]

16. An electrically driven hoist lifts a mass of 2.5 tons through a vertical height
of 80 ft in 1 min. The efficiencies of the hoist and the motor are 70%
and 80%, respectively. Calculate (a) the output horsepower of the motor,
(b) the input current to the motor if the supply is 220 V d.c. and (c) the
cost of electrical energy consumed in a month of 30 days if the hoist
operates for 4 h each day and electrical energy is 1.5 cents per kWh.

[Ans. (a) 19.4 h.p, (b) 82.2 A, (c) $32.58]

APPENDIX

MATHEMATICAL FORMULAE

Indices and logarithms

$$a^0 = 1$$

$$a^{1/2} = \sqrt{a}$$

$$a^{3/2} = \sqrt{a^3}$$

$$a^{m/n} = \sqrt[n]{a^m}$$

$$a^{-m} = 1/a^m$$

$$a^m \times a^n = a^{m+n}$$

$$a^m/a^n = a^{m-n}$$

$$(a^n)^m = a^{mn}$$

$$\log a + \log b = \log ab$$

$$\log a - \log b = \log (a/b)$$

$$\log a^n = n \log a$$

$$\log_a y = \log_{10} y / \log_{10} a$$

$$\log_e y \approx 2.3026 \log_{10} y$$

Products and factors

$$(a + b)^2 = a^2 + 2ab + b^2$$

$$(a - b)^2 = a^2 - 2ab + b^2$$

$$a^2 - b^2 = (a + b)(a - b)$$

$$(a \pm b)^3 = a^3 \pm 3a^2b + 3ab^2 \pm b^3$$

$$a^3 + b^3 = (a + b)(a^2 - ab + b^2)$$

$$a^3 - b^3 = (a - b)(a^2 + ab + b^2)$$

Quadratic equation

If $\quad ax^2 + bx + c = 0$

then $\qquad x = \dfrac{-b \pm \sqrt{(b^2 - 4ac)}}{2a}$

If $\quad b^2 > 4ac$, the roots are real and unequal

$b^2 = 4ac$, the roots are real and equal

$b^2 < 4ac$, the roots are complex

Sum of roots $= {}^{-b}\!/a$; $\qquad$ product of roots $= {}^{c}\!/a$

Arithemetical progression

$$S = a + (a + d) + (a + 2d) + \ldots$$

n^{th} term $\quad \ell = a + (n - 1)d$

Sum of n terms, $\quad S = ({}^{1}\!/2)n\,(a + \ell)$

$$= ({}^{1}\!/2)n\,\{2a + (n - 1)d\}$$

where a is the first term of the arithmetical progression of n terms

ℓ is the last term

d is the difference between successive terms

Geometrical progression

$$S = a + ar + ar^2 + \ldots$$

n^{th} term $\quad = ar^{n-1}$

Sum of n terms $\quad = \dfrac{a(1 - r^n)}{1 - r}$

Sum to infinity $\quad = {}^{a}\!/(1 - r) \ \text{ if } r < 1$

Binomial series

$$(1 + x)^n = 1 + nx + \dfrac{n(n - 1)}{1 \cdot 2} x^2 + \dfrac{n(n - 1)(n - 2)}{1 \cdot 2 \cdot 3} x^3 + \ldots$$

The series ends with x^n if n is a positive integer; for all other values of n, it is an infinite series and numerically x must be less than 1.

Exponential series

$$e^x = 1 + x + \dfrac{x^2}{2!} + \dfrac{x^3}{3!} + \ldots \text{ for all values of } x$$

Trigonometrical identities

$$\sin(90^\circ - \theta) \;=\; \cos\theta \;=\; \sin(90^\circ + \theta) \tag{1}$$

$$\cos(90^\circ - \theta) \;=\; \sin\theta \;=\; -\cos(90^\circ + \theta) \tag{2}$$

$$\tan(90^\circ - \theta) \;=\; \cot\theta \;=\; -\tan(90^\circ + \theta) \tag{3}$$

$$\sin^2\theta + \cos^2\theta \;=\; 1 \tag{4}$$

$$1 + \tan^2\theta \;=\; \sec^2\theta \tag{5}$$

$$1 + \cot^2\theta \;=\; \mathrm{cosec}^2\theta \tag{6}$$

$$\sin(A+B) \;=\; \sin A \cos B + \cos A \sin B \tag{7}$$

$$\sin(A-B) \;=\; \sin A \cos B - \cos A \sin B \tag{8}$$

$$\cos(A+B) \;=\; \cos A \cos B - \sin A \sin B \tag{9}$$

$$\cos(A-B) \;=\; \cos A \cos B + \sin A \sin B \tag{10}$$

$$\frac{\sin\theta}{\cos\theta} \;=\; \tan\theta \tag{11}$$

$$\tan(A+B) \;=\; \frac{\tan A + \tan B}{1 - \tan A \tan B} \tag{12}$$

$$(\tan A - B) \;=\; \frac{\tan A \tan B}{1 + \tan A \tan B} \tag{13}$$

Replacing B by A in (7) and (9), we obtain the following:

In (7)
$$\sin 2A \;=\; \sin(A+A)$$
$$\;=\; \sin A \cos A + \cos A \sin A$$

i.e.
$$\sin 2A \;=\; 2\sin A \cos A \tag{14}$$

In (9)
$$\cos 2A \;=\; \cos A \cos A - \sin A \sin A$$

i.e.
$$\cos 2A \;=\; \cos^2 A - \sin^2 A \tag{15}$$

Similarly, in (12) we obtain

$$\tan 2A \;=\; \frac{2\tan A}{1 - \tan^2 A} \tag{16}$$

(contd - trigonometrical identities)

Substituting from (4), $\sin^2 A = 1 - \cos^2 A$ or $\cos^2 A = 1 - \sin^2 A$

in (15), we obtain

$$\cos 2A = 2 \cos^2 A - 1 \tag{17}$$

$$\text{and} \quad \cos 2A = 1 - 2 \sin^2 A \tag{18}$$

From (17) $\qquad 2 \cos^2 A = 1 + \cos 2A$

$$\therefore \quad \cos^2 A = \tfrac{1}{2}(1 + \cos 2A) \tag{19}$$

and from (18) $\qquad 2 \sin^2 A = 1 - \cos 2A$

$$\therefore \quad \sin^2 A = \tfrac{1}{2}(1 - \cos 2A) \tag{20}$$

$$\sin A + \sin B = 2 \sin \left(\frac{A+B}{2}\right) \cos \left(\frac{A-B}{2}\right) \tag{21}$$

$$\sin A - \sin B = 2 \cos \left(\frac{A+B}{2}\right) \sin \left(\frac{A-B}{2}\right) \tag{22}$$

$$\cos A + \cos B = 2 \cos \left(\frac{A+B}{2}\right) \cos \left(\frac{A-B}{2}\right) \tag{23}$$

$$\cos A - \cos B = 2 \sin \left(\frac{A+B}{2}\right) \sin \left(\frac{B-A}{2}\right) \tag{24}$$

$$2 \sin A \cos B = \sin (A + B) + \sin (A - B) \tag{25}$$

$$2 \cos A \sin B = \sin (A + B) - \sin (A - B) \tag{26}$$

$$2 \cos A \cos B = \cos(A + B) + \cos (A - B) \tag{27}$$

$$2 \sin A \sin B = \cos (A - B) - \cos (A + B) \tag{28}$$

Triangle formulae

The sine rule: $\qquad a/\sin A = b/\sin B = c/\sin C$

The cosine rule: $\qquad a^2 = b^2 + c^2 - 2bc \cos A$

$$b^2 = a^2 + c^2 - 2ac \cos B$$

$$c^2 = a^2 + b^2 - 2ab \cos C$$

Areas

Triangle

$$\text{Area} = \tfrac{1}{2}\,ab\,\sin C = \sqrt{s[(s-a)(s-b)(s-c)]}$$

where a, b and c are the sides of the triangle

and $\qquad s = \tfrac{1}{2}(a+b+c)$

Sector of circle

$$\text{Area} = \tfrac{1}{2}\,r^2\,\theta$$

Ellipse

$$\text{Area} = \pi ab$$

Quadrilateral

Area $= \tfrac{1}{2}$ product of diagonals x sine of angle (ϕ) between them

Curved surface of cone

$$\text{Area} = \pi r \ell \quad \text{or} \quad \left(\tfrac{\pi d\ell}{2}\right)$$

Curved surface of cylinder

$$\text{Area} = 2\pi rh \quad \text{or} \quad (\pi dh)$$

Curved surface of frustum of cone

$$\text{Area} = \pi(r_1 + r_2)\,\ell$$

Surface of sphere

$$\text{Area} = 4\pi r^2 \quad \text{or} \quad (\pi d^2)$$

Surface of zone of sphere

Area $= 2\pi r$ x distance between plane ends

Irregular surface – Simpson's rule

Divide the base into an even number of equal parts of width w and erect ordinates at each point of division. Number the ordinates, finishing up with an odd number.

Area $\approx \left(^1/3\right)w$ x [sum of first and last ordinates + 2 x the sum of odd ordinates + 4 x sum of even ordinates]

Volumes

Cone

$$\text{Volume} = \tfrac{1}{3} \times \pi r^2 h \quad \text{or} \quad (\pi d^2 h / 12)$$

Cylinder

$$\text{Volume} = \pi r^2 h \quad \text{or} \quad (\pi d^2 h / 4)$$

Frustum of pyramid

$$\text{Volume} = (\tfrac{1}{3})h \, [A_1 + \sqrt{(A_1 A_2)} + A_2]$$

Frustum of cone

$$\text{Volume} = (\tfrac{1}{3})\pi h \, (r_1^2 + r_1 r_2 + r_2^2)$$

Prism

$$\text{Volume} = \text{area of base x perpendicular distance between parallel ends}$$

Pyramid

$$\text{Volume} = (\tfrac{1}{3}) \text{ area of base x perpendicular height}$$

Sphere

$$\text{Volume} = (\tfrac{4}{3})\pi r^3 \quad \text{or} \quad (\pi d^3 / 6)$$

Segment (or cap or zone) of sphere

$$\text{Volume} = (\tfrac{1}{6})\pi h \, \{3 \, (r_1^2 + r_2^2) + h^2\}$$

Sector of sphere

$$\text{Volume} = (\tfrac{2}{3})\pi r^2 h$$

Note: h is the distance between parallel plane faces and r_1 and r_2 are the radii of the faces.

<u>Refer to Fig. A</u>

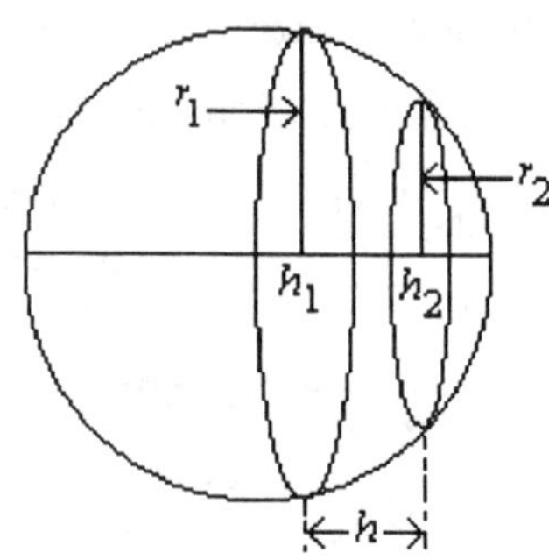

Fig. A

Differential calculus

Some Common Derivatives

$y = ax^n$	$dy/dx = anx^{n-1}$
$y = 1/x$	$dy/dx = -1/x^2$
$y = \sqrt{x}$	$dy/dx = 1/2\sqrt{x}$
$y = \sin x$	$dy/dx = \cos x$
$y = \cos x$	$dy/dx = -\sin x$
$y = \tan x$	$dy/dx = \sec^2 x$
$y = \operatorname{cosec} x$	$dy/dx = -\operatorname{cosec} x \cot x$
$y = \sec x$	$dy/dx = \sec x \tan x$
$y = a\,e^{bx}$	$dy/dx = ab\,e^{bx}$
$y = a\,e^{-bx}$	$dy/dx = -ab\,e^{-bx}$
$y = \log_e ax$	$dy/dx = 1/x$
$y = \sinh x$	$dy/dx = \cosh x$
$y = \cosh x$	$dy/dx = \sinh x$
$y = \tanh x$	$dy/dx = \operatorname{sech}^2 x$
$y = \operatorname{cosech} x$	$dy/dx = -\operatorname{cosech} x \coth x$
$y = \operatorname{sech} x$	$dy/dx = -\operatorname{sech} x \tanh x$
$y = \coth x$	$dy/dx = -\operatorname{cosech}^2 x$
$y = \sin^{-1}(x/a)$	$dy/dx = 1/\sqrt{(a^2 - x^2)}$
$y = \cos^{-1}(x/a)$	$dy/dx = -1/\sqrt{(a^2 - x^2)}$
$y = \tan^{-1}(x/a)$	$dy/dx = a/(a^2 + x^2)$
$y = a^x$	$dy/dx = a^x \log_e a$
$y = \log_a x$	$dy/dx = 1/(x \log_e a)$

Maxima and minima

At a maximum dy/dx changes sign from + to −

At a minimum dy/dx changes sign from − to +

Alternatively

If $\quad dy/dx = 0$ and d^2y/dx^2 is negative, y is a maximum

If $\quad dy/dx = 0$ and d^2y/dx^2 is positive, y is a minimum

Rules for differentiation

Product Rule

If u and v are functions of x

$$d(uv)/dx = u\,(dv/dx) + v\,(du/dx)$$

Quotient Rule

If u and v are functions of x

$$d/dx(u/v) = \frac{v\,(du/dx) - u\,(dv/dx)}{v^2}$$

Function of a function

Based on the identity $dy/dx = (dy/du) \times (du/dx)$

$$dx/dy = \frac{1}{dy/dx}$$

Logarithmic functions (logarithms to base _e_)

Some standard integrals

Logarithms to the base of _e_ are called _natural logarithms_ or _Napierian logarithms_. In calculus and in all more advanced mathematics it is customary always to use natural logarithms, and where no base is indicated it will be understood that the base is _e_.

(i) If $y = \log (ax + b)$, we can write

$$y = \log u, \quad \text{where } u = (ax + b)$$

Then $dy/du = 1/u$ and $du/dx = a$

$$\therefore \quad dy/dx = (dy/du) \times (du/dx) = (1/u) \times a = a/u$$

$$= a/(ax + b)$$

$$\therefore \quad \int [1/(ax + b)]dx = (1/a) \log (ax + b) + C$$

(ii) If $y = \log \sin x$, then

$$y = \log u, \quad \text{where } u = \sin x$$

$$\therefore \quad dy/du = 1/u \text{ and } du/dx = \cos x$$

$$\therefore \quad dy/dx = (dy/du) \times (du/dx) = (1/u) \cos x$$

$$= \frac{\cos x}{u}$$

$$= (\cos x / \sin x) = \cot x$$

$$\therefore \quad \int \cot x \, dx = \log \sin x + C$$

(iii) Similarly,

If $y = \log \cos x$, then by writing $\log \cos x$ as $\log u$, where $u = \cos x$, we have

$$d/dx(\log \cos x) = d/dx (\log u) = d/du (\log u) \times du/dx$$

$$= (1/u) \times (- \sin x)$$

$$= - \frac{\sin x}{\cos x} = - \tan x$$

$$\therefore \quad \int \tan x \, dx = - \log \cos x + C$$

Note: $- \log \cos x = \log (1/\cos x) = \log \sec x + C$

Integration

Some standard forms

(In the following integrals the constant of integration is omitted)

$$\int x^n\, dx = \frac{x^{n+1}}{n+1} \qquad (n \neq -1)$$

$$\int \frac{dx}{x} = \log_e x$$

$$\int a^x\, dx = \frac{a^x}{\log_e a}$$

$$\int e^x\, dx = e^x$$

$$\int \sin x\, dx = -\cos x$$

$$\int \cos x\, dx = \sin x$$

$$\int \tan x\, dx = \log_e \sec x$$

$$\int \sec^2 x\, dx = \tan x$$

$$\int \sec x\, dx = \log_e (\sec x + \tan x) = \log_e \tan \left(\frac{\pi}{4} + \frac{x}{2}\right)$$

$$\int \operatorname{cosec} x\, dx = -\log_e (\operatorname{cosec} x + \cot x) = \log_e \tan \frac{x}{2}$$

$$\int \cot x\, dx = \log_e \sin x$$

$$\int \sinh x\, dx = \cosh x$$

$$\int \cosh x\, dx = \sinh x$$

$$\int \tanh x\, dx = \log_e \cosh x$$

$$\int \coth x\, dx = \log_e \sinh x$$

Complex numbers

$$\text{If} \quad x + jy = r \angle \phi$$

$$\text{then} \quad x = r \cos \phi$$

$$\text{and} \quad y = r \sin \phi$$

$$r = \sqrt{(x^2 + y^2)}$$

$$\phi = \tan^{-1}(y/x)$$

Multiplication

$$r_1 \angle \phi_1 \times r_2 \angle \phi_2 = r_1 r_2 \angle (\phi_1 + \phi_2)$$

Division

$$\frac{r_1 \angle \phi_1}{r_2 \angle \phi_2} = (r_1/r_2) \angle (\phi_1 + \phi_2)$$